SOLAR HEATING AND COOLING OF COMMERCIAL BUILDINGS

Research project

BEng (Hons) Civil Engineering

Division of Civil & Building Services Engineering

School of The Built Environment & Architecture

Bogdan Ciocoiu

April 2016

Abstract

Solar heating and cooling technologies are an eco-friendly technique which uses energy from the Sun to heat water which later is used for domestic purposes or for controlling the air temperature within rooms or dwellings. Between 40-60% of electricity costs would be required for water heating and air heating. By using solar heating and cooling at its full potential, an average of up to 50% from monthly bills could be saved when calculating the standard on 12 months' basis.

Solar heating and cooling can be used entirely by itself to produce hot water, if the setup benefits from favourable climate and optimal hardware components, i.e. appropriate solar panels correctly positioned and heat storage systems. Further costs can be saved by storing the heated water during the night in separate tanks. By doing so, hot water will remain available in the early hours of the morning, making the boiler not required to be turned on.

Apart from the financial impact, solar heating and cooling benefit the environment by reducing the greenhouse effect proportionally, by not having to produce gases which would have been used for heating large quantities of water. Various international bodies aim for a considerable reduction of pollution by 2050, and as discussed in the next pages, solar heating and cooling play an essential part within this macro initiative.

This report looks at how solar heating and cooling can be integrated within UK schools to heat water using solar heating and cooling mechanisms, to decrease the amount of energy consumption used for boilers and to assess technical options and recommend solar heating and cooling setups to maximise efficiency, given UK weather patterns and historical exposure to Sun.

Table of Contents

Introduction

Energy management, usage optimisation and seeking integration with renewable energy sources will be a crucial objective for coming generations. The need to reduce levels of greenhouse gas ("GHG") emissions to stop the global temperature decreasing is also a challenge; hence sectors mainly running on combustion engines are slowly moving towards electric motors (cars, trains, motorbikes).

The constant increase in population and the need to satisfy a more significant demand collaborated with the need to control the pollution and reduce the volume of greenhouse gas, determines high profile international bodies such as the European Union and International Energy Agency to stimulate the market to promote eco-friendly energy sources.

Principles of solar heating and cooling

Solar Heating and Cooling ("SHC") technologies first became available around the 1960s within Australia, Japan and Israel.

Within the early 1970s, International Energy Agency was the first considerably large institution which started to promote and stimulate the market to spread the word about the new low-carbon emission energy sources, to facilitate implementation in other parts of the globe. SHC[1] is supported by international representative bodies including the European Commission - Industry, Research and Energy Department (European Parliament, 2016), ECREEE[2] (Regional Center for Renewable Energy and Energy Efficiency, 2010), ECI[3], ISES[4], GORD[5] and RCREEE[6].

Solar heating and cooling is a technique specialized in capturing heat radiation transmitted by the Sun and storing it within dedicated environments, in which temperature conservations are facilitated at most and later serving it to the consumer's application by delivering hot water, heating the air, heating the radiators, underneath the floor, water within the pool.

The technique has been implemented within both private and public sectors, by both consumers and businesses. For being technology as opposed to just a product, it comes in multiple forms based on the size and volume of deployments, coverage, environment, the amount of water which needs storing, duration of exposure to Sun, exposure to wind and multiple other factors.

Research project scope

The research project aims to assess the efficiency of solar heating and cooling for commercial buildings concerning schools and dwellings used for academic purposes. Reports look into legal and critical requirements to be fulfilled in schools (Gov.uk, 2016), i.e. temperature control and comfort, to be satisfied, to enable dwellings to be used as schools/for academic purposes.

The project also assesses the climate and environmental factors such as exposure to the Sun and the elements, extreme weather, methods for amplifying the amount of captured energy and adjacent aspects which may impact the performance of the solar heating and cooling technology. The project uses secondary sources of data and integrates both qualitative and quantitative methods.

---

[1] Solar Heating and Cooling
[2] ECOWAS Regional Centre for Renewable Energy and Energy Efficiency
[3] European Copper Institute
[4] International Solar Energy Society
[5] Gulf Organization for Research & Development
[6] Regional Center for Renewable Energy and Energy Efficiency

Authorities and institutions supporting solar heating and cooling

At present, solar heating and cooling are supported by multiple agencies and non-government organisations. The most significant authority, the European Union, implemented a programme in 2006 to control energy supply and environmental issues for pollution (Commission, 2006). The plan was proposed by Andris Piebalgs – EU Energy Commissioner. The strategy was introduced on 19/Oct/2006, and it aims to reach a 20% energy reduction by 2020.

International Energy Agency ("IEA") is a crucial member creating awareness campaigns to support solar heating and cooling through direct involvement at the governmental level and regular events, interacting with members of the public and private sector. International Solar Energy Society ("ISES") organises ISES Solar World Congress meetings around the world to spread awareness.

Some of the locations are key large cities such as Freiburg or Kassel - Germany, Paris – France, Beijing – China or Sydney – Australia, where they join forces with smaller organisations such as Australian Solar Energy Society (AuSES) to advance and promote both the development and the deployment of solar heating and cooling technologies.

Solar Thermal Technology Panel ("STTP") is another child organisation supporting IEA's initiative. Experts formed it within the solar heating and cooling field, and it aims to answer market-specific critical questions about solar heating and cooling and to stimulate manufacturers to implement new features to address market needs.

Research methodology

This evaluation relies on historical weather reports to obtain accurate information about the duration of daylight throughout the years. It is taking into account several aspects of the research.

Firstly, the story will look into solar heating and cooling efficiency, benefits of integration, equipment performance, anatomy, serviceability, popularity and it also includes financial indicators. It will then look at several representative study cases and aim to identify the extent to which solar heating and cooling were implemented.

The report aims to produce estimative figures of equipment, i.e. solar panels required for schools by assessing existing study cases and merging with known data about the number of pupils and hot water consumption. It will also look into existing programs run by the UK government to facilitate access to SHC technology, i.e. grants and loans, particularly local council involvement with schools and relevant local authorities.

Literature Review

Schools considered as commercial buildings

According to the Cambridge Business English Dictionary, 'commercial building' is a *'building used for business activities'* (Press, 2016). 'Business activities' is defined as *'any activity that is engaged in for the primary purpose of making profit'* (Investopedia, 2016). The education activity has a cost whether it is supported primarily from a loan or subsidised by the government; therefore, schools do perform a commercial operation.

It has been decided to focus the paper on schools to gain a better understanding of the standards, the requirements, the benefits and the negative aspects of implementing solar heating and cooling technologies in such environments.

Energy required for heating and cooling purposes

The importance of heat and the need to identify current energy sources for heating is reflected within 2009 International Energy Agency report (International Energy Agency, 2009) which states that 47% of the total energy used at any moment in time is being used for heating purposes, while only 17% is used for electricity and 27% for transport. According to Carbon Trust, over 50% of the energy consumed within schools is being used for heating purposes (Carbon Trust, 2004).

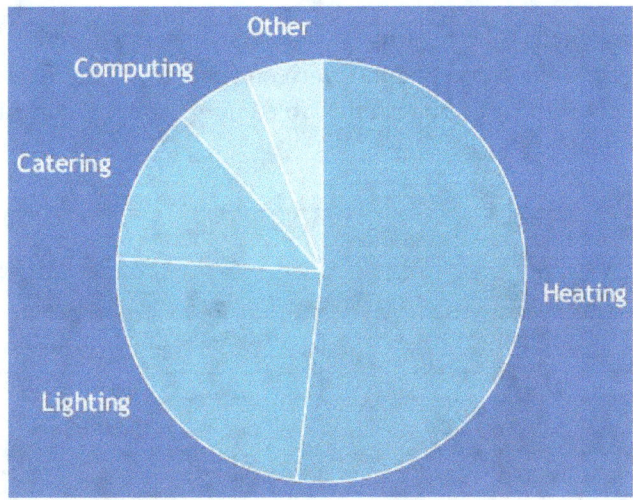

Breakdown of energy use in average school (Carbon Trust, 2004)

History of solar heating and cooling

Solar Heating and Cooling mechanisms were first deployed on a large scale during the 1960s. Ten years later, the International Energy Agency ("IEA") were organizing worldwide events to advertise the technology (International Energy Agency, 2016).

Between the 1970s and 1980s, working closely with Global Solar Thermal Energy Council (Global Solar Thermal Energy Council, 2010), IEA got the message across to Europe, North America and the United States about the importance of solar heating technology. At the end of 2012, 0.38 billion square meters were used worldwide to place solar panel collectors within a total of 58 countries, representing 4.4 billion people, which in effect consists of 63% of the overall world's population.

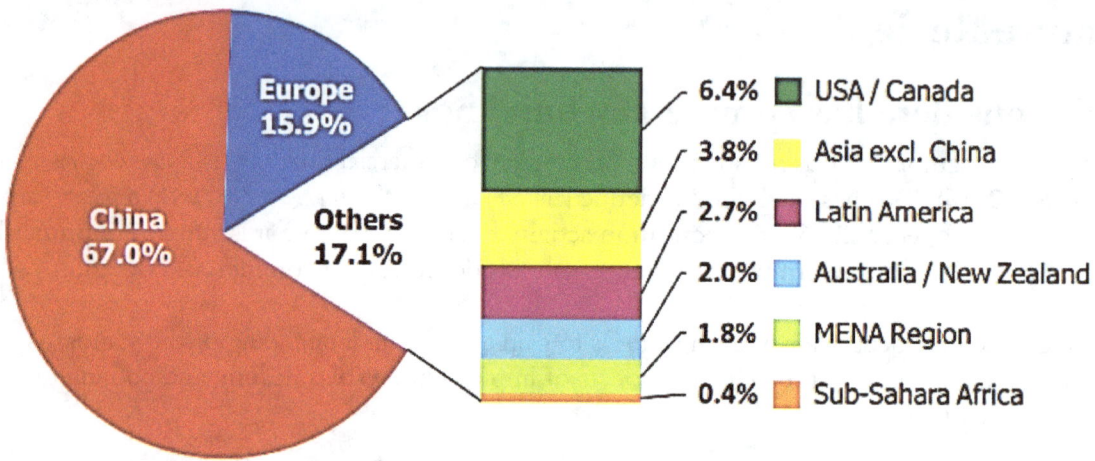

Sub-Sahara Africa:	Mozambique, Namibia, South Africa, Zimbabwe
Asia excluding China:	India, Japan, Korea South, Taiwan, Thailand
Latin America:	Brazil, Chile, Mexico, Uruguay
Europe:	EU 28, Albania, Macedonia, Norway, Switzerland, Russia, Turkey
MENA Region:	Israel, Jordan, Lebanon, Morocco, Tunisia

Share of the total installed capacity in operation – 2012 (Mauthner & Weiss, 2012)

During 2013 90% of the areas where solar hearing was actively in service were China, Germany, India, Japan, Austria, Brazil, Portugal, Mexico and Spain. Fourteen years from the first records of solar heating and cooling, it became the second most popular renewable energy source, contributing to the global energy, second to wind power (Mauthner & Weiss, 2014).

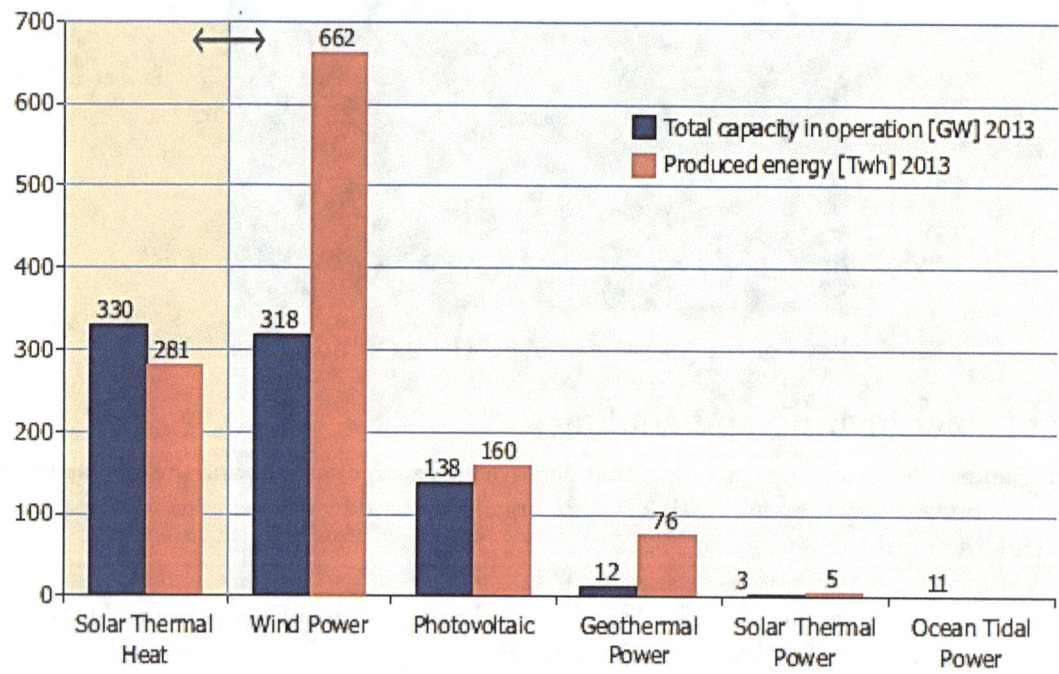

Total cap. GWel[7], GWth[8], prod. energy/y. TWhel[9], TWhth[10] - 2013 (Mauthner & Weiss, 2012)

---

[7] Gigawatts-electricity
[8] Gigawatts-thermal
[9] Terawatts-electricity
[10] Terawatts-thermal

The vision of solar heating and cooling

International Energy Agency estimates an increase in the number of application types that use solar radiation as a result of constant technological improvement, industry training, specific marketing, future useful workshops, a decrease of manufacturing cost for solar heating and cooling equipment and ultimately increase within accessibility.

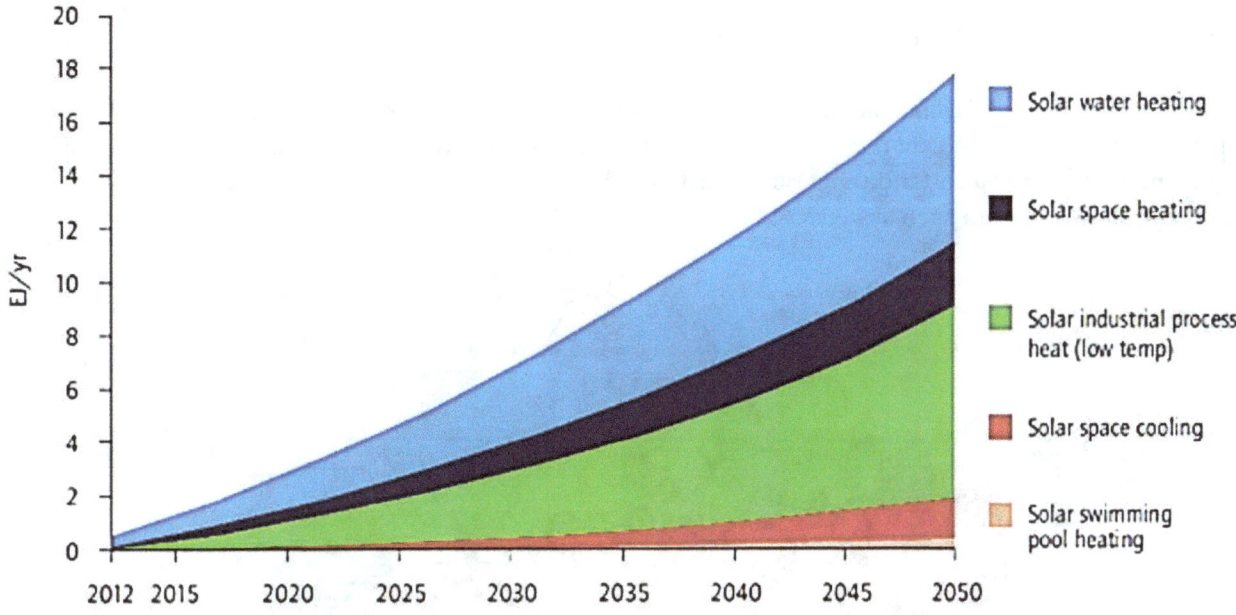

Vision for Solar Heating and Cooling (Exajoule/year) (International Energy Agency, 2012)

The report suggests that apart from the existing types of applications which use solar heating and cooling (i.e. solar water heating, solar space heating, solar industrial heating or solar swimming pool heating), many other systems will be developed through constant variation and technological breakthrough.

	Water collectors		Total
	FPC	ETC	m^2
UK	75,600	29,600	105,200
Worldwide	10,433,875	47,186,980	60,106,338
Ratios UK/worldwide	0.72%	0.06%	0.18%

New collectors installed during 2010 – area covered m2 – UK (Mauthner & Weiss, 2012)

	Water collectors		Total
	FPC	ETC	m^2
UK	72,953	18,826	97,376
Worldwide	10,249,865	56,223,156	68,813,468
Ratios UK/worldwide	0.71%	0.03%	0.14%

New collectors installed during 2011 – area covered m2 – UK (Mauthner & Weiss, 2012)

	Water collectors		Total
	FPC	ETC	m^2

UK	490,097	160,301	664,398
Worldwide	94,450,147	208,918,828	336,257,996
Ratios UK/worldwide	0.52%	0.08%	0.20%

Total collector installed by end of 2011 – area covered m2 – UK (Mauthner & Weiss, 2012)

To reach the expected objective for 2050 (as indicated by the International Energy Agency), a long term international policy must be agreed and implemented across the countries to promote and assist with the implementation of solar heating and cooling within schools. Projects such as Sustainable Building Envelope Demonstration ("SBED") must be initiated between members of the private sector and local authorities of governmental bodies (Williams & Manager, 2016).

The need for such is to facilitate access to loans and European funds, increase the level of confidence from the private sector and stimulate the market into creating new jobs and initiatives within long term solar heating and cooling projects, and trigger interest from the private sector of potential customers. They would be willing to set up such systems for their use.

Sustainable Building Envelope Demonstration (SBED) - project scope (Brown, et al., 2013)

Sustainable Building Envelope Demonstration benefits from £1.8 million obtained from the Welsh government. The budget was allocated from the European Regional Development Fund. Other objectives constitute from learning frameworks and facilitating access to information to deliver an understanding about solar heating and cooling to those who seek to learn more about it to facilitate potential conversion – this is done via awareness campaigns, training, online documentation and workshops.

The goal is for solar heating and cooling to become accessible and affordable through low-interest rate loans and unique programs which will allow schools to run long term plans for integrating such equipment. This objective can be achieved by merging efforts from multiple countries along with local or central government support and coordination.

Key benefits of using solar heating and cooling

Solar Heating and Cooling runs on a variety of technologies and offers a variety of applications; however, the most common technology refers to the solar hot water system (Solar Energy Industries Association, 2011). The key benefits of this include:

- Provides free hot water and heating - with an excellent technical solution it can also store and deliver hot water over 24+ hours;
- Can be positioned virtually anywhere as long as panels are exposed to the Sun-on-roof, in-roof, over walls, on canopies or over the ground surface using inclined supports;
- Extends boiler lifespan by avoiding usage while reserves of hot water are still available and within desired temperatures;
- Non-polluting — carbon-free (except for production and transportation); several systems involve Sun tracking mechanisms which use electricity to power motors to rotate panels;
- Government grants reduced 5% VAT and council engagement; as shown within the report, Oxfordshire Council engaged with all local public schools to initiate an unprecedented project, aiming to install up to 5000 solar panels;
- Can be combined with photovoltaics in highly efficient cogeneration schemes to convert heat into energy which can be used, stored or returned to the national grid;
- Quiet—few or no moving parts – no fans or motors required within essential setups; advanced systems may involve mechanical movement outside the building;
- Mature technology—multiple manufacturers and service providers – available worldwide – initiated within the 1960s;
- Excellent return on investment for long term planning – autonomy of over 25 years – value stands in the ability to design a system to suit the climate and structural options to return maximum efficiency;
- Operating costs are near-zero;
- Franz Mauthner and Warner Weiss suggest that during 2013, the number of jobs relating to solar heating and cooling, including both installation and maintenance roles, worldwide, were around 460,000. This list shows the financial impact produced by solar heating and cooling technology.

Alternative eco-friendly sources of energy

Within Europe, three critical sources of energy are being used: thermal energy, biomass energy and geothermal energy.

Thermal energy	Is generated by the movement of particles in an object – action which is being stimulated by excessive heat (Schools, 2016).
Biomass chemical energy	Form in which sunlight is being stored in materials such as wood, straw or manure (ESDB, 2016).
Geothermal energy	It is obtained from heat generated from within the Earth through hot water or hot rock.

Alternative eco-friendly energy sources

As above, alternative technologies are expensive to implement, require space to an industrial scale and also require specialized resources to set up and monitor the production of energy. None of the options above would be as easy and efficient to install as solar heating and cooling and certainly not feasible within schools.

Technical assessment

Solar heating

Solar heating principles play a crucial role as it delivers hot water while producing low quantities of greenhouse gas (GHG) emissions, mitigating against climate change.

For being non-restrictive (from the integration perspective) it can be easily implemented within various both domestic and commercial environments. The basic principle behind solar heating and cooling refers to objects left in direct contact to the Sun, which for being exposed to solar radiation absorb part of it, converting it into thermal energy.

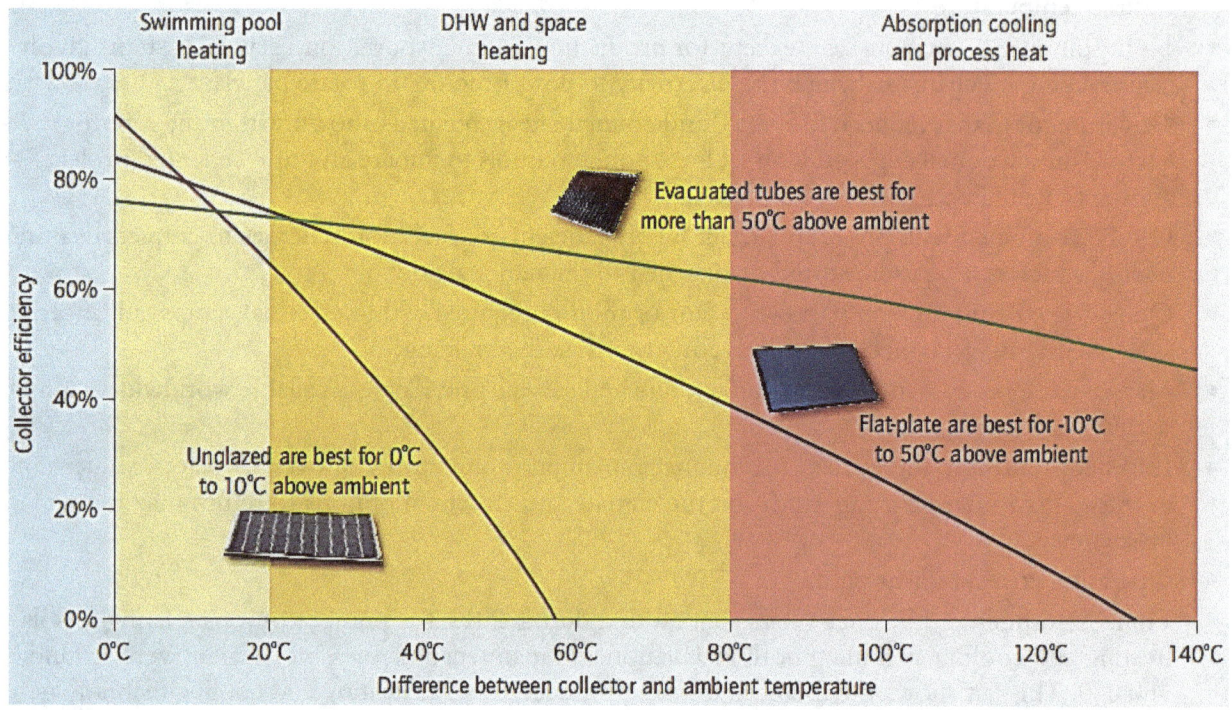

Outside temperature with collector efficiency (International Energy Agency, 2012)

Once captured, the thermal energy will then be stored or circulated and used for convenience. The fundamental principle behind solar heating and cooling starts from the energy absorption phenomenon, to which any object positioned in the direct contact with the Sun, is subjected to solar radiation, will absorb part of it and have it converted into thermal energy.

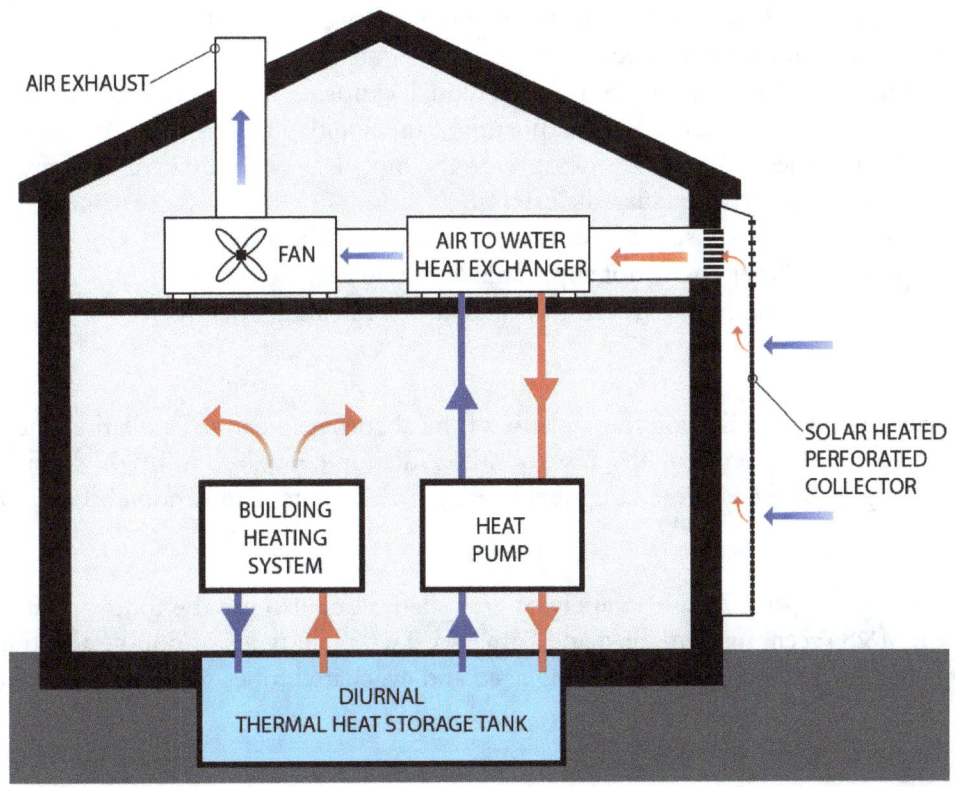

Core principle of solar heating and cooling (Brown, et al., 2013)

Solar thermal mechanisms work on a variety of temperature levels, from 25-30°C available within swimming pool collectors, all the way up to 250°C within concentrated solar technology systems.

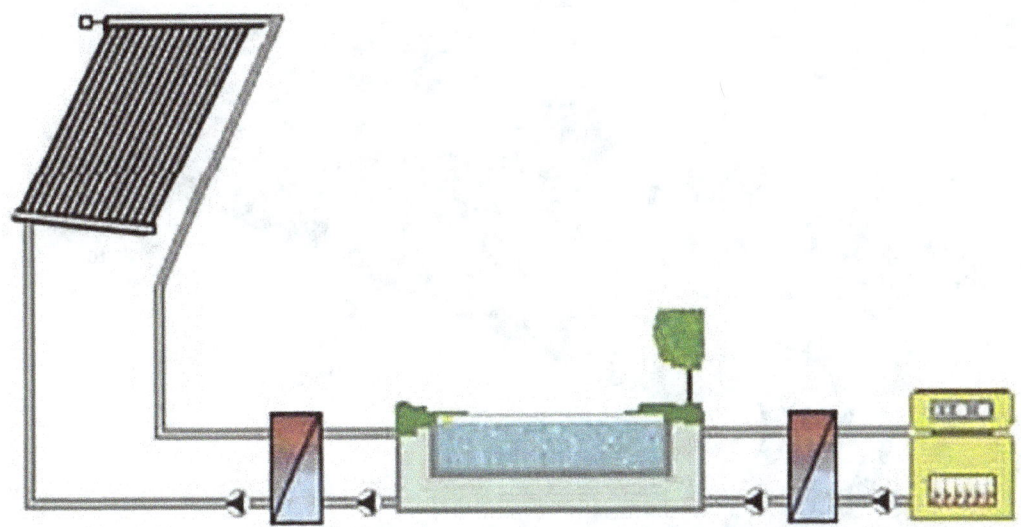

Swimming pool solar heating system (Clean Green Energy Zone, 2005)

All designs have four elements in common:

Absorber	The absorber collects incoming near-infrared and visible solar radiation. It can be distinguished through its dark colour as dark colours absorb heat.

Selective absorbers	Collects incoming solar radiation and also releases infrared radiation to maximize the amount of heat absorbed.
Circuit	Heat transfer fluid flows through a tunnel, exposing the liquid to heat. This process takes place to maximize the exposure of the liquid, which is being subjected to heat. Multiple designs have an absorber on the outside area of the hydraulic circuit, for the same reason to maximize the effects.
Housing	Reduces energy loses caused by absorber and the heating circuit. Housing is similar to a protective layer and is not applicable with unglazed collectors.

Common parts from principle solar heating systems

Colour of receptors

Colour is a critical element of indicating the capacity of the absorber to capture radiation and have it converted to heat; specifically, dark colours absorb more radiation compared to bright colours. The civil engineering industry goes as far as producing steel coating with advanced radiation absorption, including light colours.

Transpired solar collectors ("TSC") use a variety of grey shades but also use the following four colours: Chocolate brown, M&S green, Linden green and Sargasso. Two main types of non-concentrating thermal collectors: flat plate which could be glazed or unglazed and evacuated tubes which can use direct flow tubes or heat pipe tubes.

Flat plate collectors

Flat plates are absorber or collector cells which form the solar panel and have the ability to capture heat transmitted from the Sun. The plates are positioned within small housing cases usually made from aluminium or steel for protection. The plates are insulated and protected by a layer of vacuum, to reduce thermal loss.

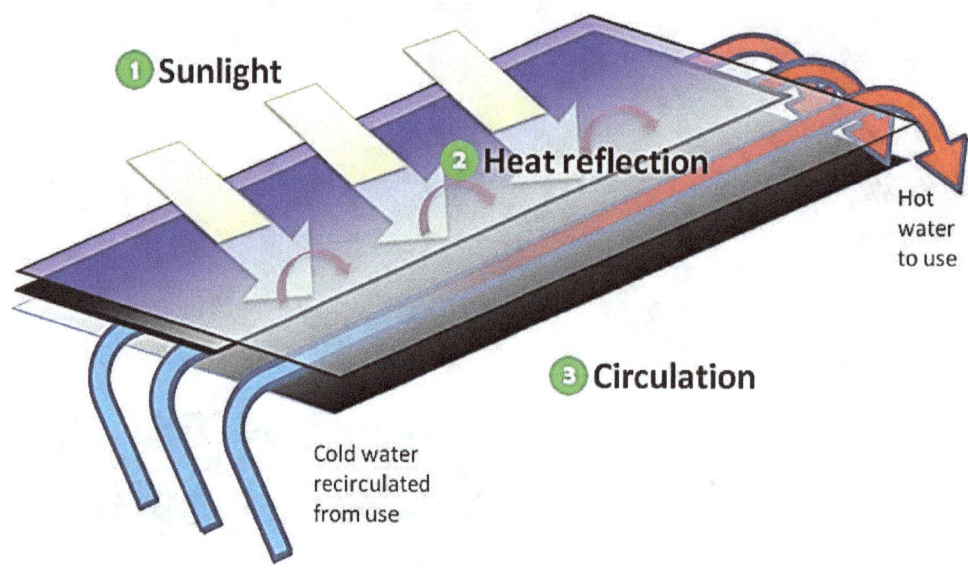

Flat plate collector showing area exposed to Sun (Go Green Heat Solutions, 2005)

One of the advantages for flat plate collectors is that they can capture radiation from all directions from the absorber side, not only the radiation transmitted perpendicularly, therefore tracking the Sun through mechanical means is not required. Constant water circulation assures uniform heating across the entire system. For circulating systems, hot water will heat across the tubing path, i.e. across a room, floor or house.

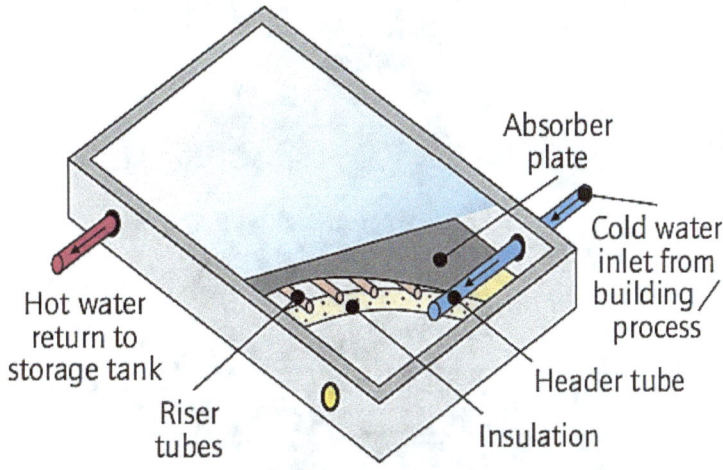

Flat plate collector showing internal insulation layers (International Energy Agency, 2012)

	Daily solar radiation average (MJ/m²) – 12mo average calculated	
	New Mexico	UK
Flat plate collector - Tracking on 2 axes	31	19.5
Flat plate collector - Fixed	23	15
Concentrator - Tracking on 2 axes	26.5	14

Concentrator vs flat plate; New Mexico vs UK; fixed or adjustable (Power of The Sun, 2010)

Glazed flat plate collectors

Glazed flat plates, as the name suggests have a transparent layer on top of the flat panels. The coating acts as an isolator increasing the working temperature for the plates but also protecting the plates against losing heat should the wind blow towards the solar panels.

Glazed solar panels are recommended for recycling heated air or recirculating heated air systems, particularly for buildings or for space heating. Collectors have an energy collecting surface on the side exposed to the Sun.

Collector with a layer of glazing on front side (Build it Solar, 2005)

Increasing the number of sheets or layers which are covering the collectors, will also increase permitted temperature to which the collectors can be exposed. A good example consists of the triple glazed collectors who were designed to operate at extreme temperatures (Power of The Sun, 2010) while double glazed or single glazed will cover lower temperature bands.

Glazing layer supported over the collectors (Build it Solar, 2005)

The above figure shows a glazing layer attached over collectors (Builditsolar, 2016). The critical difference between flat plate collectors and single layer glazed collectors in terms of reliability and resistance to the wind is that for non-glazed panels, wind can disperse the heat captured by the absorbers. Glazing has a deflective purpose, and given that schools are positioned within both urban and rural areas, the chances of having exposure to the wind are considerably high, therefore having at least one glazing layer will benefit the system.

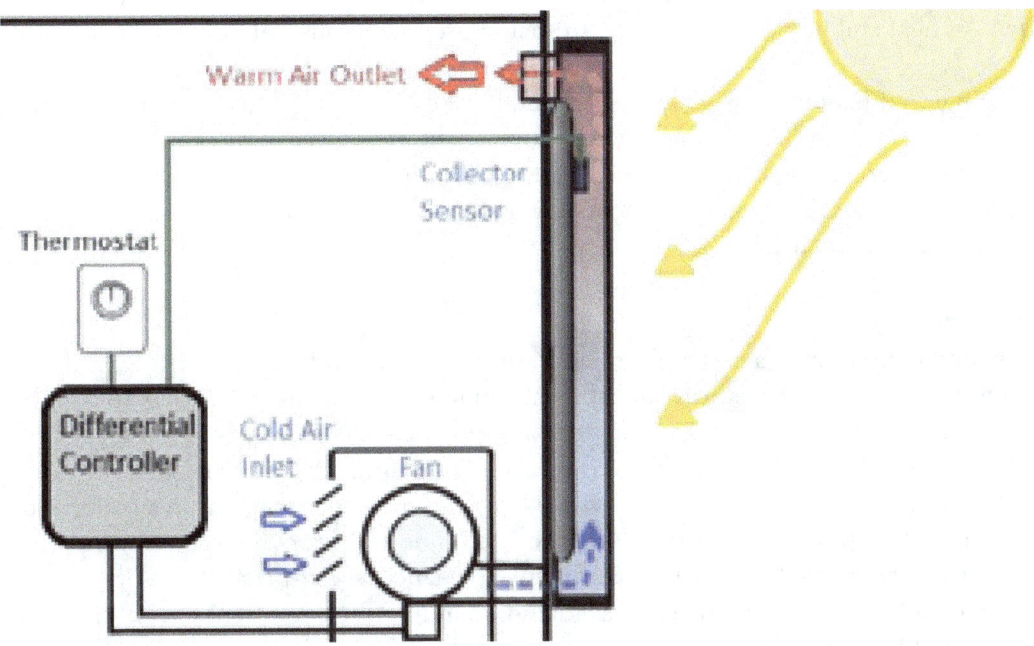

Glazed air collectors – diagram (Clean Energy Resource Teams, 2012)

The above system shows an air collector mounted on the entire height of a wall on the side of the building. The benefits of using such systems reflect the fact that the entire height of the room can be heated without requiring a fan or similar. It should be noted that such panels can easily be positioned on all floors of a building and through piping work, the entire structure can be heated, not just the side exposed to the Sun.

Glazed air collectors - commercial deployment (Ener Concept, 2004)

The image above shows an example building which was covered entirely in solar panels on the side exposed to the Sun.

Unglazed flat-plate collectors

Unglazed flat plates absorb heat at a lower temperature, making them friendlier to the UK climate; however, they are more sensitive to wind and snow. The metallic or plastic absorbers have no covering and are mainly used for heat ambient of outside air – primarily used for industrial or agriculture purposes.

The market for unglazed flat plate collectors also includes applications for heating swimming pools and fish farming systems. As a result of the fact that there is no insulation, such systems can work at temperatures as low as 30°C. Since there is no glazing, the plates absorb a more significant amount of heat from the Sun, however since there is no insulation a more substantial portion of heat is also lost due to wind or if the outside temperature is cold (Darling, 2000).

Evacuated tubes

The tubes are manufactured from a material similar to durable glass, with high levels of transparency. The key stands within the tube's wall thickness, and standard practices allow manufacturers to produce pipes with a wall thickness of 1.8mm which prove efficient and resistance when exposed to rain and windy weather (Aprocus, 2010). Within the tube, there is a vacuum, assuring isolation and therefore, low heat loss. Small insulated evacuated tubes perform better where load requires high temperatures and the ambient temperatures are low. If deployed within the UK, evacuated tube collectors have a well-insulated design; therefore their performance would be better than the performance of flat-place collectors due to the predominance of clouds which can induce low irradiation conditions quite often. Given the above, evacuated tubes could represent the right solution for schools within UK schools, as a result of their resistance to unfavourable cloudy weather.

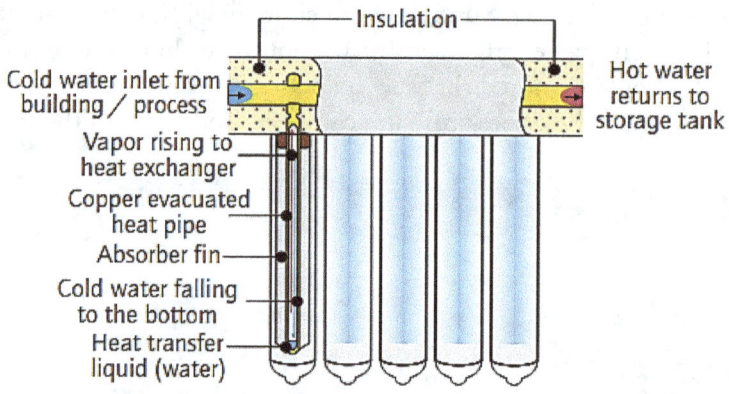

Evacuated tube collector (International Energy Agency, 2012)

Direct flow tubes

Another common type is called the Sydney tube, which has two main subcategories: Twin-glass or thermos Flask tube. Direct flow tubes use as the main principle the circulation of water within the container by exposing it to Sun. Within the tube, there is a vacuum layer which has the function of isolating and protecting against heat loss.

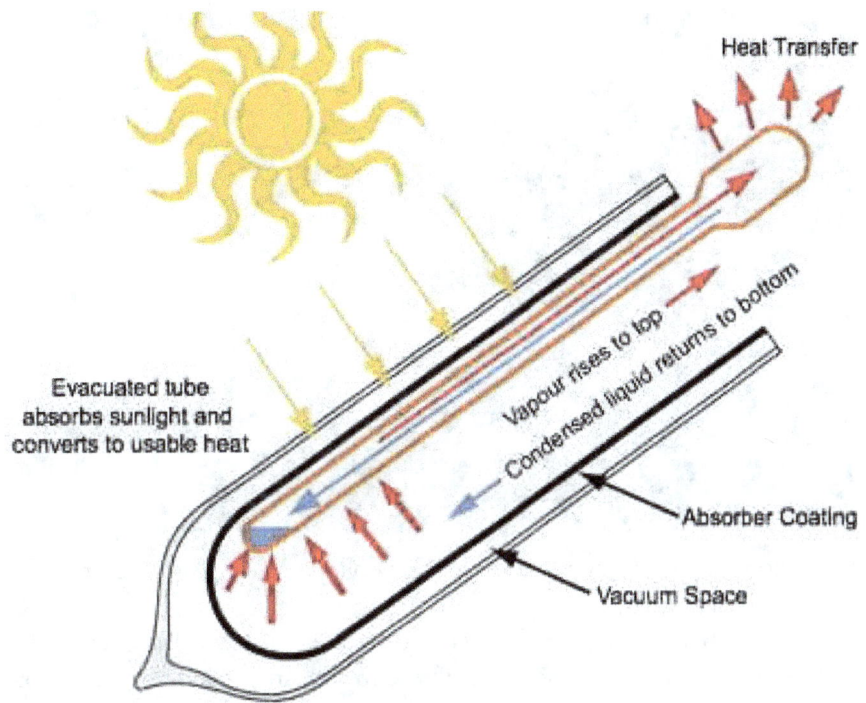

Direct flow tubes - Twin glass (Apricus, 2003)

A similar principle with the vacuum bottles is that they maintain the temperature of drinks, i.e. keep the coffee warm. The most common ones contain two separate tubes, one inner, one outer with a vacuum between to avoid facilitating temperature transfer (Marken, 2009).

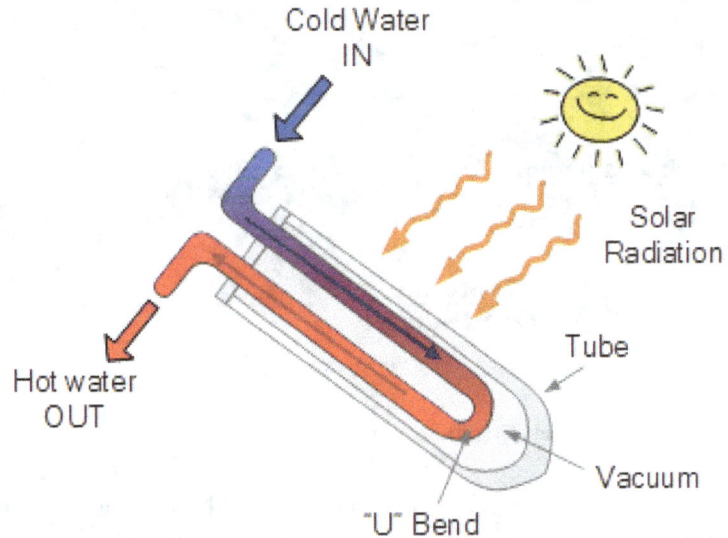

Direct flow tubes - Flask tube (Alternative Energy, 2004)

Another common type includes a "U bend" which has a smaller tube inside which transfers liquid. The side which exposes the liquid to the Sun heats, while the second half allows the water to exit.

Direct flow tubes - Evacuated tube solar collectors (In Light Solar, 2001)

Relatively easy to attach and replace if damaged, direct flow tubes represent an excellent portable solution ideal for portable/removable set-ups. Small schools where mobilising and demobilising is required frequently; the direct flow tubes can be fitted within portable supports such as the one above, which will facilitate quick deployment.

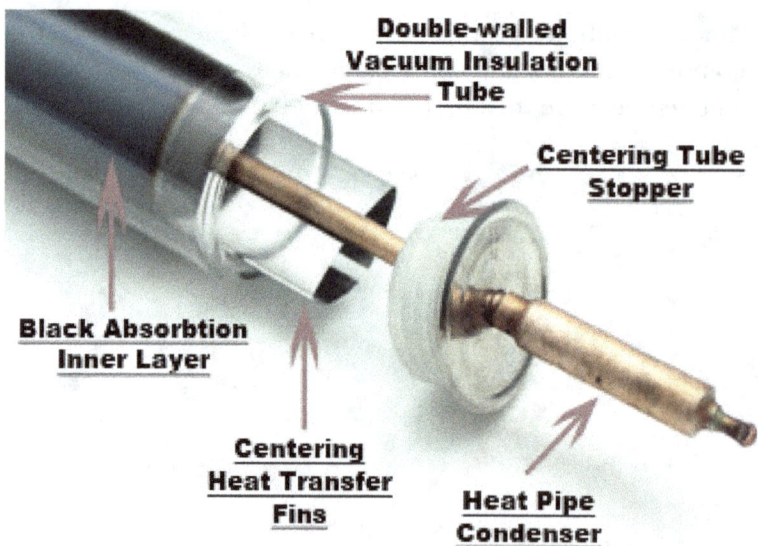

Direct flow tubes - Evacuated tube solar collectors – contents (Solar Tubs, 2005)

Evacuated tube solar collectors produce excellent efficiency in bright sunshine. For ease, evacuated tube collectors have a colour indicator often through a barium getter. The purpose of the feature is to highlight by changing its colour to white for when vacuum will get compromised (Solar Panales Plus, 2014).

Heat pipe tubes
Heat pipe tubes follow a relatively similar principle with the one used by direct flow tubes except for the fact that the fluid is exposed through a single inner tube. Heat pipe tubes also use void to protect against heat loss.

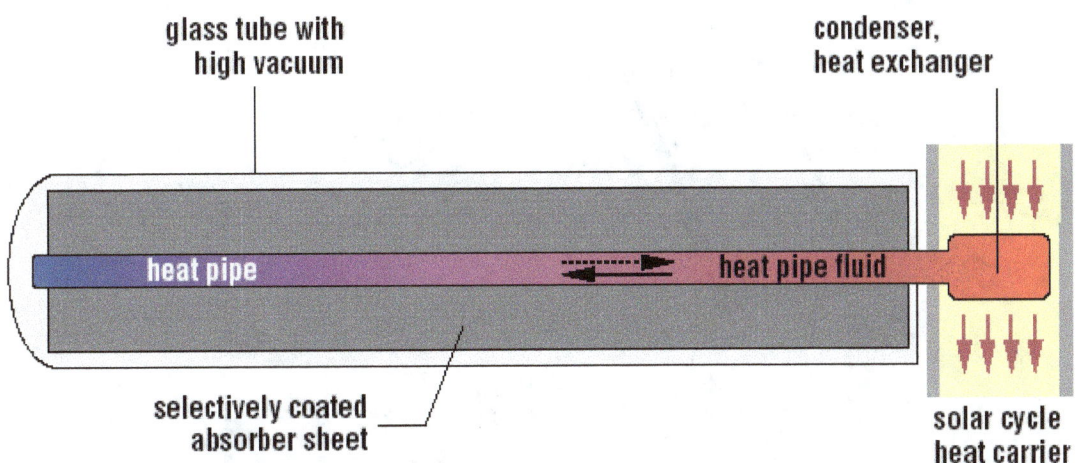

glass tube with
high vacuum

condenser,
heat exchanger

heat pipe   heat pipe fluid

selectively coated
absorber sheet

solar cycle
heat carrier

Typical heat pipe tube - evacuated tube solar collectors (Volker Quaschning, 2004)

Heat pipe tubes can operate at higher temperatures compared to flat plate tubes, mainly as a result of the width of the layer between the inner and the outer cells, which is more significant for heat pipe tubes.

Circulation system types

Thermosiphon or natural circulation system types are appropriate for environments when frost does not take place. The critical principle consists of the flow of the heated liquid, which for having a higher temperature will float over the non-heated liquid.

Thermosiphon or natural circulation system (International Energy Agency, 2012)

Within thermosiphon or natural circulation mechanisms, the collector will need to be positioned next to the heat storage. The advantages consist of not having any pumps involved and no electricity required. The disadvantage consists of the fact that the collector and the heat storage may require a considerable volume of space which may not always be available.

Pumped or forced circulation

For being equipped with pumps, such systems allow separation between the collector and heat storage. However, pipework is required to enable the liquid transfer.

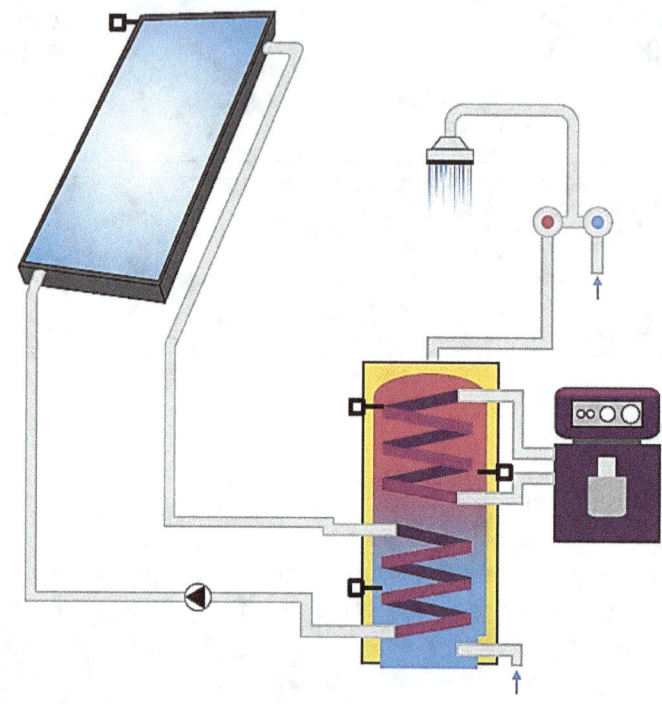

Pumped or circulation system (International Energy Agency, 2012)

These systems increase the level of complexity by introducing pumps; however, they can function within environments where low temperatures are being recorded, which is not an option for natural circulation systems. Because pumps are being used, forced circulation systems require a source of electricity; therefore, wiring and appropriate cabling are also required. Pumped technologies are particularly useful for the generation of hot water but also for facilitating space heating.

Solar cooling

The industry started to support standardization of such devices from 2007 onwards, so all accessories will fit when being installed. Since standardization started, a cost reduction of 50% has been acknowledged. Solar cooling principles refer to the conversion of energy initially received from the Sun into the cold through driving a thermal cooling machine with thermal energy generated with solar thermal collectors.

As of 2011	About 750 solar cooling systems were installed worldwide.
As of 2013	A total number of 1050 solar cooling systems installed worldwide, from which 80% are in Europe – especially Spain, Germany and Italy.

Details about solar cooling deployment between 2011 and 2013

Solar cooling mechanisms use sorbent and refrigerant; therefore, the maintenance process will require adjustments on their levels.

No. of solar cooling installations [-]

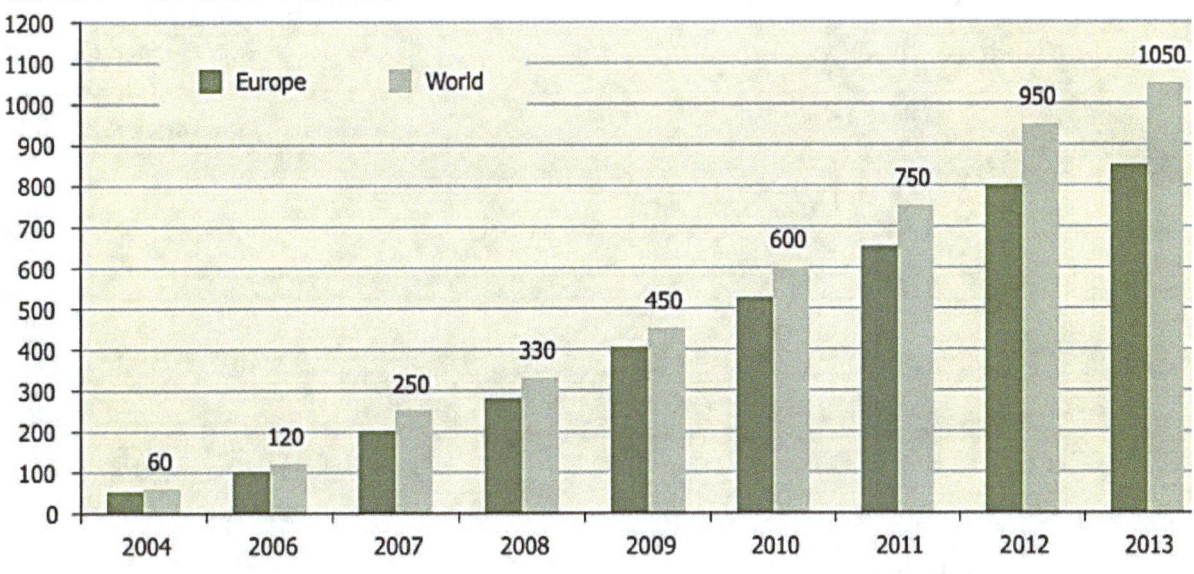

Solar cooling development trend (Mauthner & Weiss, 2012)

The technique can reduce electrically powered cooling considerably if deployed on a large scale. Thermal storage is also an option for solar cooling if adequate thermal storage is being provided.

Closed cycles

Within closed-cycle systems, the liquid is not in direct contact with the environment; however, it is circulated within a closed pipe system. Air is filtered, and its temperature is cooled by exposing it to the above piping work, then supplied to the limited space to be distributed within the room. Solar cooling closed-cycle systems use absorption chillers to remove heat from the circulating liquid. The chillers can produce the required amount of cooling using hot water delivered from solar panels (Alternative Energy, 2004).

Chilled water produced by these systems can be supplied to any air conditioning hardware. Absorption chillers are extremely common within the solar cooling technology as a result of the fact that they can work efficiently even at lower temperatures such as 55°C with a COP of 0.6. Average temperatures practised would be between 70°C and 100°C and a COP[11] of 0.7.; these are often called single-effect absorption chillers. Given the frequency of cloudy weather, high temperatures may not always be continuously reached, therefore the heat absorption system becomes relevant within the efficiency of the cooling system.

---

[11] Coefficient of performance - thermal efficiency of the chillers

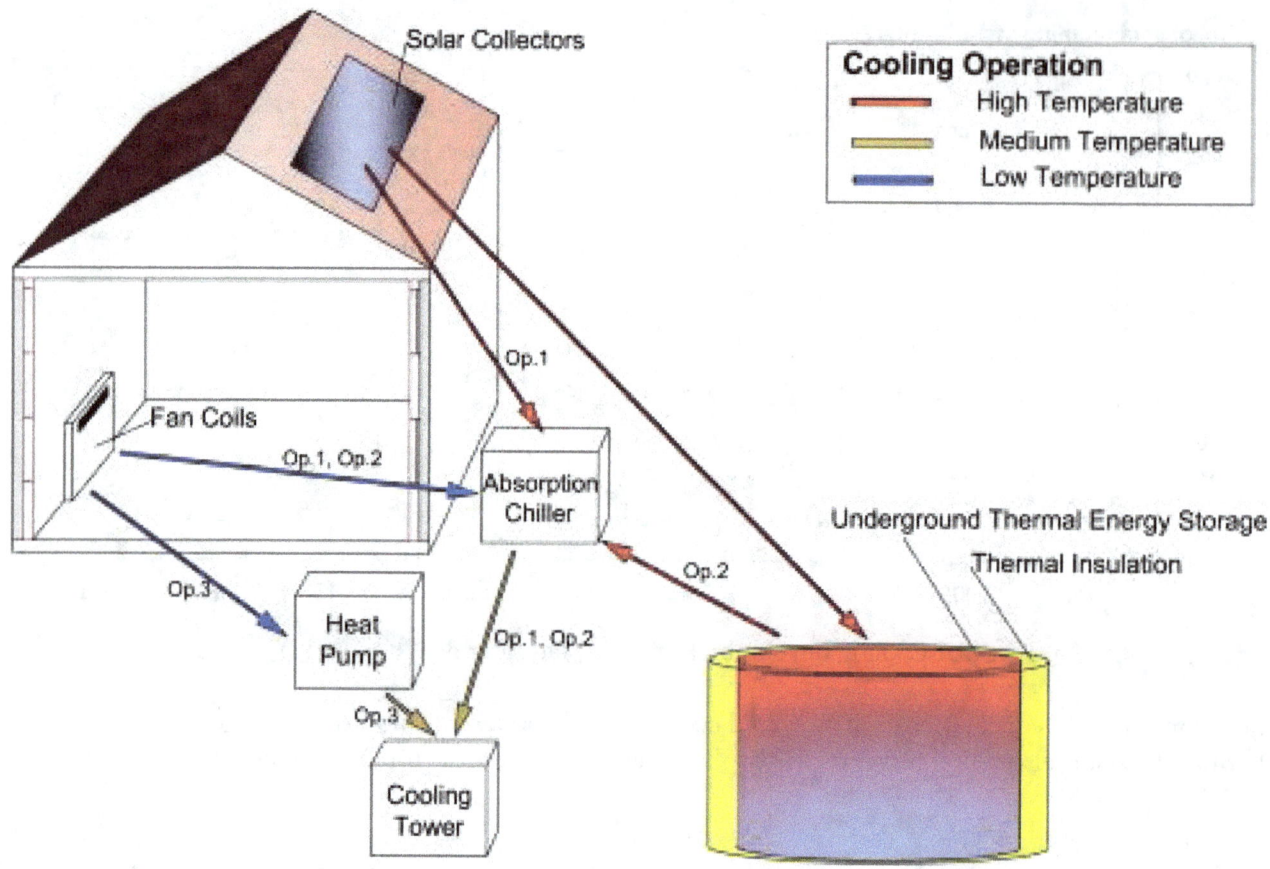

Absorption chillers and heat pump – SHC system (The Renewable Energy Centre, 2005)

Another type includes the 'double effect' absorption chillers. The second system is using two thermal generators connected as series. For being serially connected, they increase the COP to 1.1-1.2. The driving temperatures are also high; they can reach between 150°C and 180°C. The market is relatively new for triple-effect cycles. However, they can achieve COP values of 1.6-1.9 under working temperatures up to 200°C and 250°C. For double effect and triple effect closed-cycle systems, there's also a question of applicability to the UK cloudy weather. The capacity of the driving temperature influences the design of the collector – a higher capacity requires a better collector.

Open cycles

The treatment system uses water as the refrigerant and a desiccant for the sorbent. Open cycle systems use DEC[12] mechanisms. Contrary to the closed cycles, open cycles directly produce cold air. Thermally driven open cooling cycles function on evaporative cooling and air dehumidification, through the use of desiccant or similar. This technology offers control over the humidity levels, and it also delivers well within space cooling purposes. Open cycle solar cooling is relatively new technology even for the advanced markets and for this research, precedents of deployments of such within schools could not be found.

Solar air heating

Solar air heating is a specific technique through which air circulated internally is being heated using heat from the Sun. The energy required to perform space heating is much more expensive compared to the power used for heating water for basic domestic applications. When referring to schools, providing every school has a considerable number of rooms, and that space heating needs to take place within every room, it becomes relevant that air heating systems will have a positive impact from both financial and ecological

---

[12] Desiccant evaporative cooling system

perspectives. In terms of energy demand water heating requires only 1/5 from the total amount of energy needed for space heating (Build it solar, 2010).

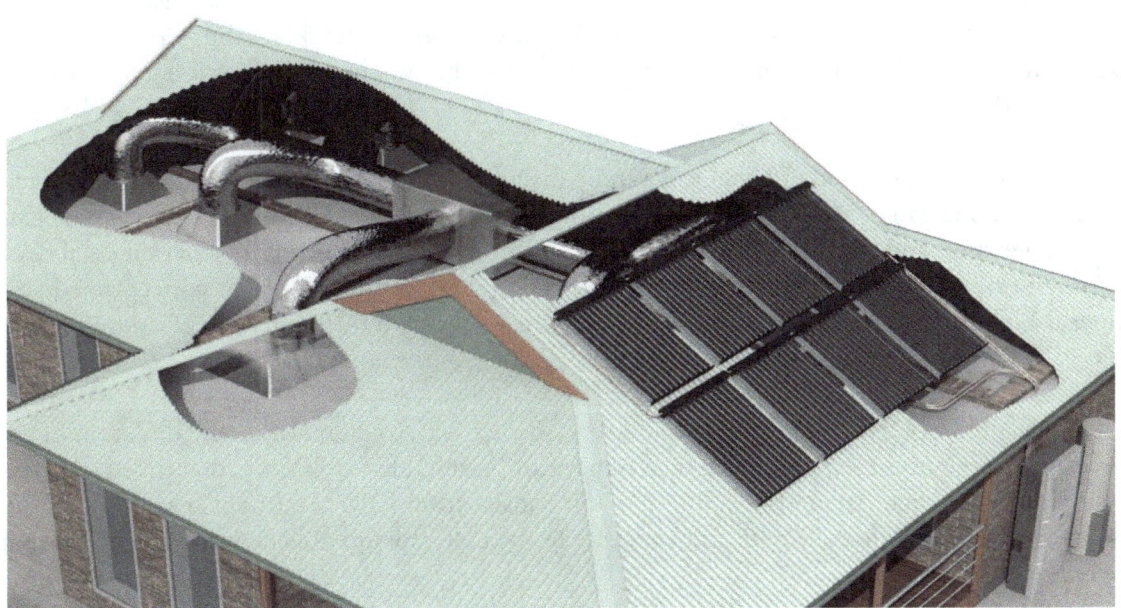

3D design showing ducting work connected to solar air heating system (Enefield, 2004)

Solar air heating systems can be integrated better and easier during the building's construction process. At maximum potential, solar air heating can reduce electric consumption with up to 20-30% from total conventional energy used for air heating. Solar air heating is being used within agriculture for crop drying but also within schools, military, industrial and residential environments. Evidence of solar air heating systems suggests it has been used for ~30 years, yet is not as well sought as solar water heating. The primary mechanism involves circulating heated air through ducts and fans into the building's ventilation system.

Panels used in solar air heating (Ecoactive, 2013)

The ability to store heated air can be designed and integrated within the system – therefore, solar air heating is fully scalable. North America has the majority of solar air heating systems currently deployed for both domestic and commercial purposes applications.

Concentrating solar systems

Concentrating solar systems amplify the efficiency of conventional solar heating systems, by concentrating Sunlight focus before reaching the absorbers. When concentrated, the volume of heat is higher than if non-concentrated, therefore more efficient once captured. If used with Sun tracking, the performance will increase even more, and this will enable further integration within the heavy industry, such as using steam turbines and producing electricity.

Furthermore, given the increased efficiency, it can be installed with CHP[13] systems. In terms of tracking, the most effective method involves movement on both the X and Y-axis. Potential issues can be caused by wind and bad weather, and because of the complexity of the system, protecting by covering will decrease the performance of the system substantially. It is also dependent on the number of concentrated systems if the number if high the investment may not be suitable for a school, but for small-sized systems, it can be an option.

The energy measured in DNI[14] – energy received if the device is perpendicular to the direction of the Sun's rays. This scenario requires clear skies, and it also may not be suitable for UK weather given the frequency of rain and cloudy weather. An interim solution could consist from CPC[15] mechanisms, a reflector used in both non-evacuated flat plate collectors and evacuated tube collectors to collect and concentrate radiation from the Sun (Edmound, 2010).

Heat storage devices

Heat storage devices are critical elements for storing heat to extend its delivery for 24-48 hours or such. Heat storing systems are scalable; however, they are dependant to the isolating technique and volume of water. They also require a considerable amount of space; however, schools usually benefit from a significant amount of space, so it will be only a logistical matter to allocate a room for such.

---

[13] Heat and power systems
[14] Direct normal irradiance
[15] Compound parabolic concentrator

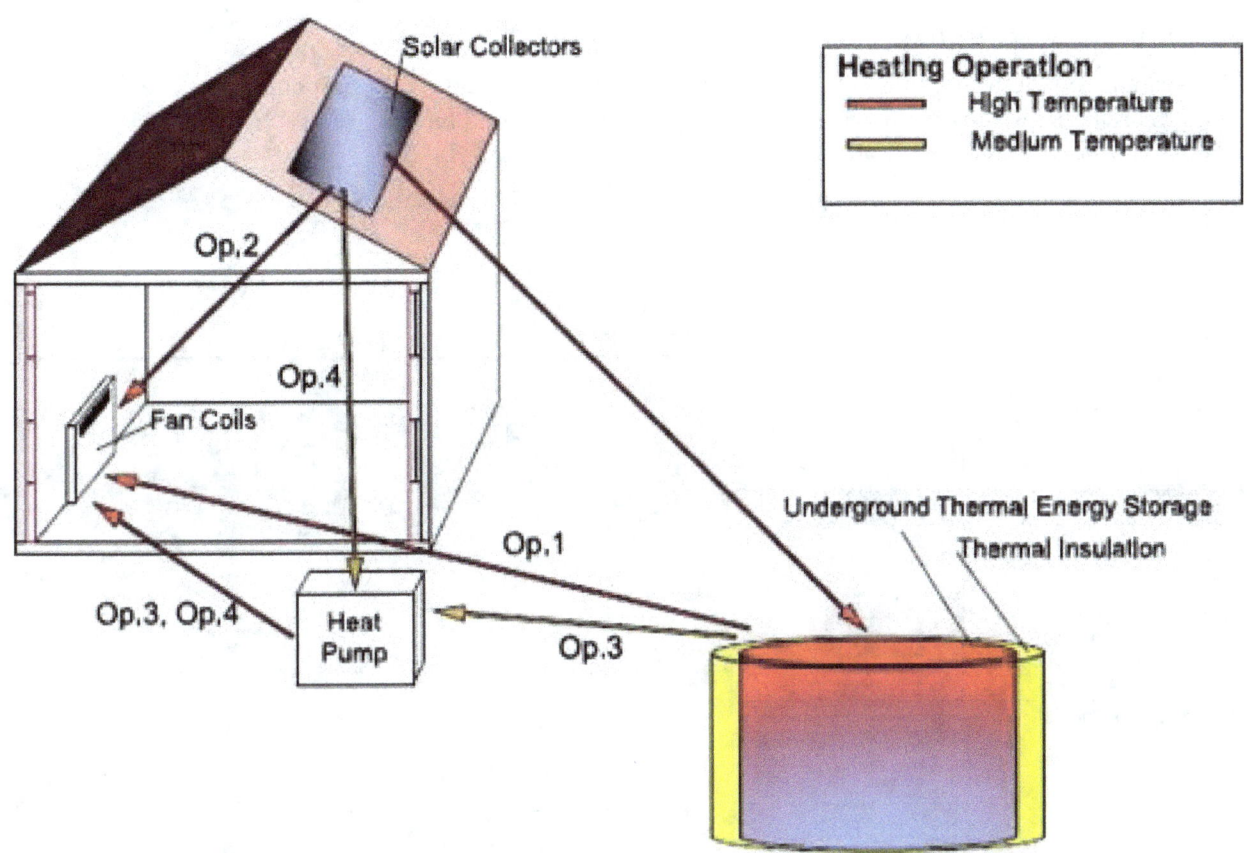

The majority of manufacturers provide solutions which should store hot water for periods of up to 24 hours however multiple systems can be installed to run in parallel as long as the number of solar panels allows such deployment.

Sensible heat storage	• High popularity therefore competitive costs • Heat can be stored at higher temperatures than 100°C • Uses water as heat storage and transfer medium
Latent heat storage	• Systems use melting or evaporation, therefore, running temperatures are higher than above • Existing passive systems are currently used for low-temperature storage to improve performance.
Sorption heat storage	• The system uses water vapour uptake • Technologies are in the development phase • Theoretical densities within the sorption system can be four times higher than the theoretical density within the water.
Thermochemical heat storage	• Uses chemical reactions to manage heat • Chemicals store heat 20 times more densely compared to water's ability to store heat • Average 8-10 times more efficient than using water systems • Still in development, being tested with salts as anhydrous.

Types of storage heat devices

Techniques working temperature ranges

Each mechanism is focused on a specific temperature range, implicitly for a particular purpose depending on the effectiveness, deployment costs, scalability, maintenance and environmental risks such as exposure to rain and snow.

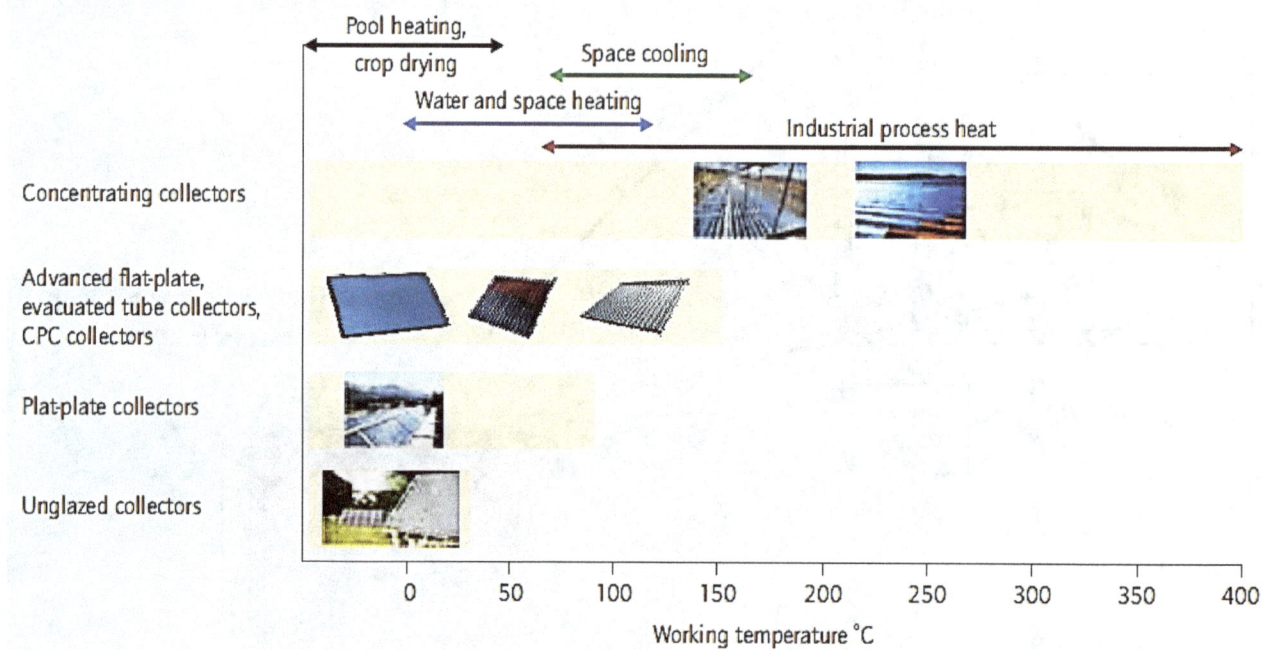

System types (y axis) vs storage temperatures (x axis) (Mauthner & Weiss, 2012)

When addressing to the above table, schools would fall under unglazed collectors and potentially flat plate collectors. For specific deployments, there is potential to implement more advanced systems which can include concentrated solar systems; however given the expensive installation and maintenance costs, it is less likely to witness such deployments within UK schools.

Performance of applications

Solar heating and cooling are currently being used within both domestic and commercial environments worldwide. From a total estimate of 78 million water-based solar working systems serving worldwide, the following distribution takes place.

78%	Used for hot water preparation in domestic environments
8%	Used for swimming pool heating for domestic use
9%	Used within considerable larger domestic applications, i.e. multifamily, however for the same internal purposes – heating water
4%	Used within combo-systems both domestic hot water and space heating
1%	Used for industrial purposes and thermally driven applications

Solar heating systems - usage distribution

Duration of daylight throughout the year in UK

Daylight duration varies based on the geographical location of the assessed point. When referring to the United Kingdom, the following durations are being recorded (Barrow, 2016).

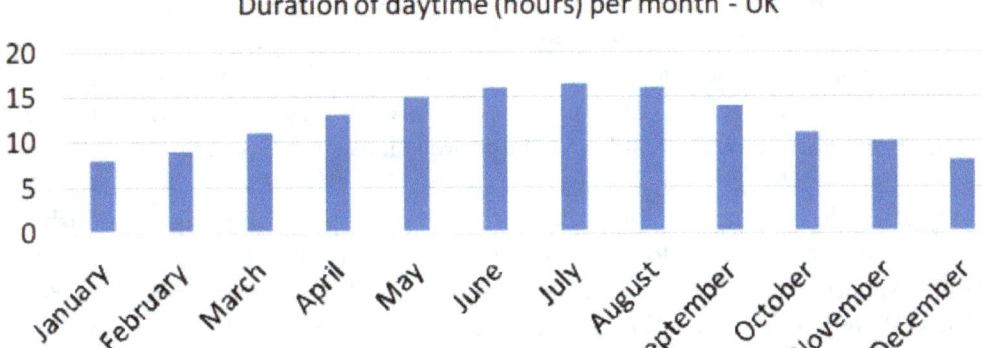

Duration of daylight in hours - UK timestamp (Barrow, 2016)

Given the critical dependency to expose solar radiation receptors to the Sun, solar heating and cooling technology is dependent on the amount of exposure to the Sun, duration of daylight and clearance to the Sun – i.e. not having cloudy weather. It's important to mention that UK schools conduct activity during all four seasons; therefore, the solar heating and cooling solution must be viable throughout the year. Given the average weather will include clouds and rain, the ideal system needs to be extremely receptive and absorb heat in less ideal circumstances and work at temperatures as low as 25°C.

Optimal tilt angle for the solar panels

The position of the solar panels has the potential to maximize or minimize the amount of exposure of the solar cells. When factoring in the additional self-rotation of the Earth around its axis, the following patterns, are likely to occur (Greenstream Publishing Limited, 2015).

Winter	Spring and Autumn	Summer
21st of Dec. Rise 58° East of due South Set 58° West of due South	21st of Mar./21st of Sep. Rise 91° East of due South Set 91° West of due South	21st of Jun. Rise 126° East of due South Set 126° West of due South

Inclination of solar panels for maximum exposure (Greenstream Publishing Limited, 2015)

For a monthly distribution, the following shows the ideal angle to capture most of the Sun's radiated heat:

Month	Jan.	Feb.	Mar.	Apr.	May	Jun.
	22°	30°	38°	46°	54°	62°
Degrees from vertical position	Jul.	Aug.	Sep.	Oct.	Nov.	Dec.
	54°	46°	38°	30°	22°	14°

Inclination of solar panels – monthly max. exposure (Greenstream Publishing Limited, 2015)

Adjusting the inclination angle can be done manually or automatically. Mechanical or digital systems which can adjust its inclination angle can be expensive; therefore some systems attempt to set a particular angle so that Sun radiation is captured perpendicularly on the solar plates and therefore the system will benefit from the highest efficiency rates. It should be mentioned that for requiring a large capacity of hot water, schools may need a large number of solar panels and if each solar board requires a mechanical system to adjust inclination, this may increase the setup costs considerably. As a result, solutions which may not require tilting to adjust the performance considerably may be more suitable for UK schools.

TSC (Transpired solar collectors) technology advises for the panel position to have an unshaded orientation within 20° of South. However, few deployments had the panels mounted on the façade of the dwellings or the roof. The only exception as of December 2013 takes place at Jaguar Land Rover UK training academy who have deployed their TSC system at 69° from a side façade (Brown, et al., 2013). The building façade perpendicular to the ground level mounting position is the standard most commonly used for the glazed air collectors usually used for the solar heating technology.

Glazed air collectors - solar heating technology (Shiftnrg, 2004)

Glazed air collectors allow for easy mounting, do not require mechanisms to adjust inclination angle, and as long as the system has exposure to the Sun, it will require low installation costs. Depending on the structure of the building, glazed air collectors could be mounted on two outside walls – an assessment would need to be done to obtain the best position judging by the exposure to the Sun.

Building Envelope Centre (SBEC) designed with cassette panels over cladding (Sbec, 2003)

Building Envelope Centre uses cassette panels to mix the technological desire to have solar panels with the aesthetic aspect of the building. Such deployments involve both design and manufacturing costs which are not likely to be allocated for small schools; however, more prominent schools with a prestigious history may look towards, such as investment.

Willmott Dixon Healthcare Campus designed with profiled cladding (Building, 2007)

Willmott Dixon Healthcare Campus was designed using lateral cladding to cover a considerable area from one side of the building. It is likely for schools to follow the same design paten given the efficiency produced by a system which covers a wide area and the potential less expensive design specific requirements.

RCT Homes Dwelling designed with tongue and groove planks (SBED, 2006)

Another example which is likely to be deployed within schools is the model implemented at RCT Homes Dwelling, similar to the one within the previous case. The technical solution is consisted of attaching non-tilting panels to schools' external walls. The efficiency is also reflected within the cleaning and maintenance processes where school maintenance can reach the solar plates using ladders and essential washing/cleaning equipment.

Domestic applications

Solar hot water heating

The majority of solar heating and domestic cooling systems are integrated to heat water. For small applications, the collector size is usually between 3-6m². The storage capacities for domestic applications start with a range between 150-300 litres. Statistics show that the total number of deployments of solar heating and domestic cooling systems in the UK is less than 0.3% compared to the total number of implementations worldwide.

	Total collector area	Collector area per system	Total number of systems	Type of system
Units	m^2	m^2		
UK	709,673	4.0	177,418	PS16
Worldwide	299,476,976	4.3	74,350,976	Mixed
%	0.24%	93.02%	0.24%	-

Solar heating and cooling deployment within UK vs. worldwide (Mauthner & Weiss, 2012)

---

[16] Pumped system

From the total area covered globally from all solar heating and cooling systems currently recorded, the United Kingdom uses 0.24%, with an area of ~709,000m². In terms of the coverage for a single panel collector, systems within the United Kingdom rely on relatively the same area with the common area for systems used worldwide. This option can suggest an average better effectiveness for UK panels compared to panels used worldwide.

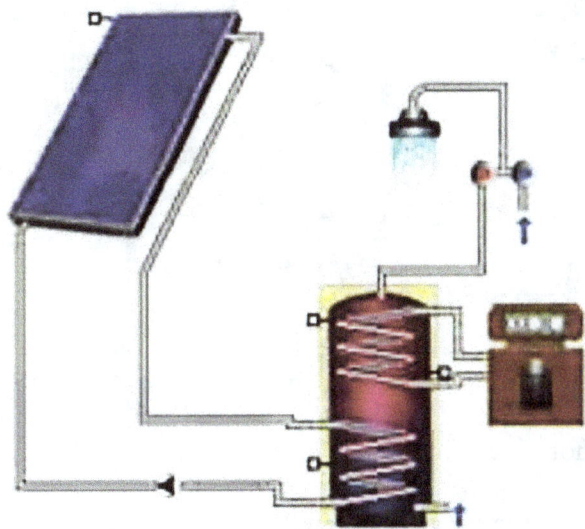

Domestic hot water pump system (BASC, 2004)

Standalone solar heating and cooling mechanisms can cover 30% from regular domestic use in average per annum. With an isolated storage tank installed, the solar heating application can cover up to 100% from the standard moderate domestic needs, if located in an appropriate climate.

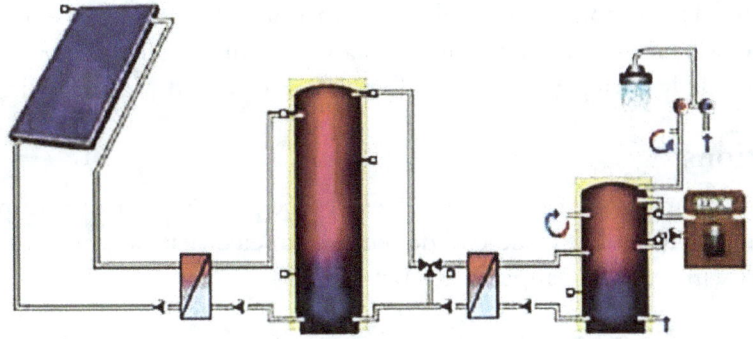

Multifamily domestic hot water pump system (BASC, 2004)

Hot water and space heating (2-in-1)
Hot water and space heating systems perform the same features as the previous type except they also heat air using a separate mechanism which subjects air to heat and after heated, it circulates the air through ducting or air ventilation independent device.

The 2-in-1 systems must have a broader exposure to the Sun to heat more water, therefore a wider area dedicated to solar panels. The ratio between the energy used for heating water and the energy used for heating space is 1/5, therefore the higher the number of solar panels dedicated for heating air, the bigger the savings towards the energy consumed.

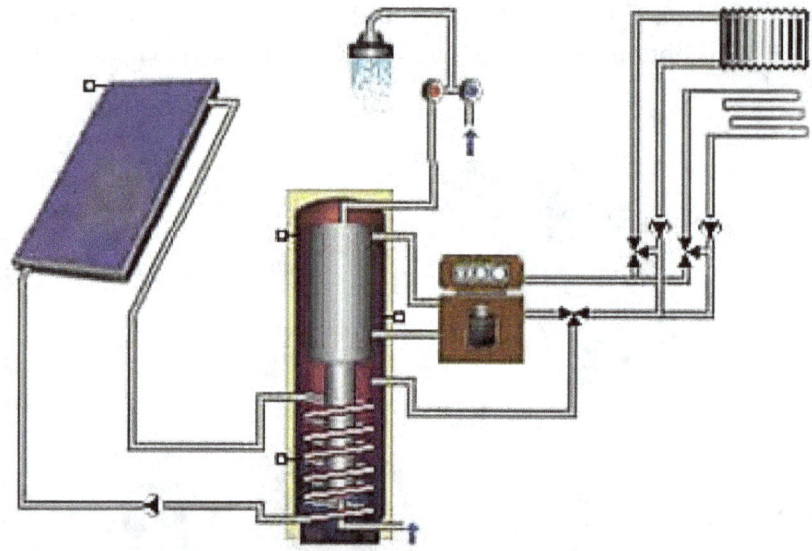

Domestic hot water and space heating – combo-system (BASC, 2004)

Within central Europe, combo-systems save between 25-30% from total fuel energy used within domestic applications of solar heating and cooling. This system needs an auxiliary energy source for a backup, given the high demand for air heating purposes and required storage.

The combo system is not as widely spread as the standard one as a result of the fact that it is slightly more expensive, and it requires ductwork or ventilation channel for air circulation. If such ductwork or air ventilation systems are not available, houses must undergo valuable works; therefore, they are less likely to be accepted by landlords. Within technically advanced countries such as Germany and Australia, combi-systems are providing up to 50% from the annual capacity of hot water required for a standard-sized household. Such combo systems are also suitable for establishments located at higher altitudes where radiation increases, particularly during September – October and March-May periods.

Industrial applications

Industrial deployment of solar heating and cooling doesn't take place within all European countries, and where it does take place, it is to reduce peak load demand from electricity suppliers. Industrial applications use systems which work with average temperatures between 30-100°C.

Commercial applications - schools

Legal requirements for temperatures in schools
The Education School Premises Regulations 1999 act provides requirements (Department for Education and Employment, 1999) along with guidance on for implementing the standards within DfEE Guidance 0029/2000, Standards for School Premises (DfES Pupil Health and Safety Team, 1998). Part IV, Section 20, Clause 2 requires the following temperatures to be satisfied by all schools within the UK:

- Areas with an average level of physical activity (i.e. classrooms) - 18°C
- Sick rooms or areas where there is less physical activity - 21°C
- Gymnasia or washrooms - 15°C

In terms of hot water, clause 22 from the same section as above requires a maximum temperature of 43°C. The Workplace (Health, Safety and Welfare) Regulations 1992 also introduce requirements about the minimum temperature for premises (UK Crown, 2013):

- HSC Approved Code of Practice—16°C

More user-friendly official documentation was released in various formats, providing the same requirements (Health and Safety Executive, 2011) assisting managers in understanding room temperature requirements.

Number of schools

As the research paper refers to the United Kingdom, the following information was extracted to show a relatively constant number of pupils attending UK public schools (Gov.uk, 2015).

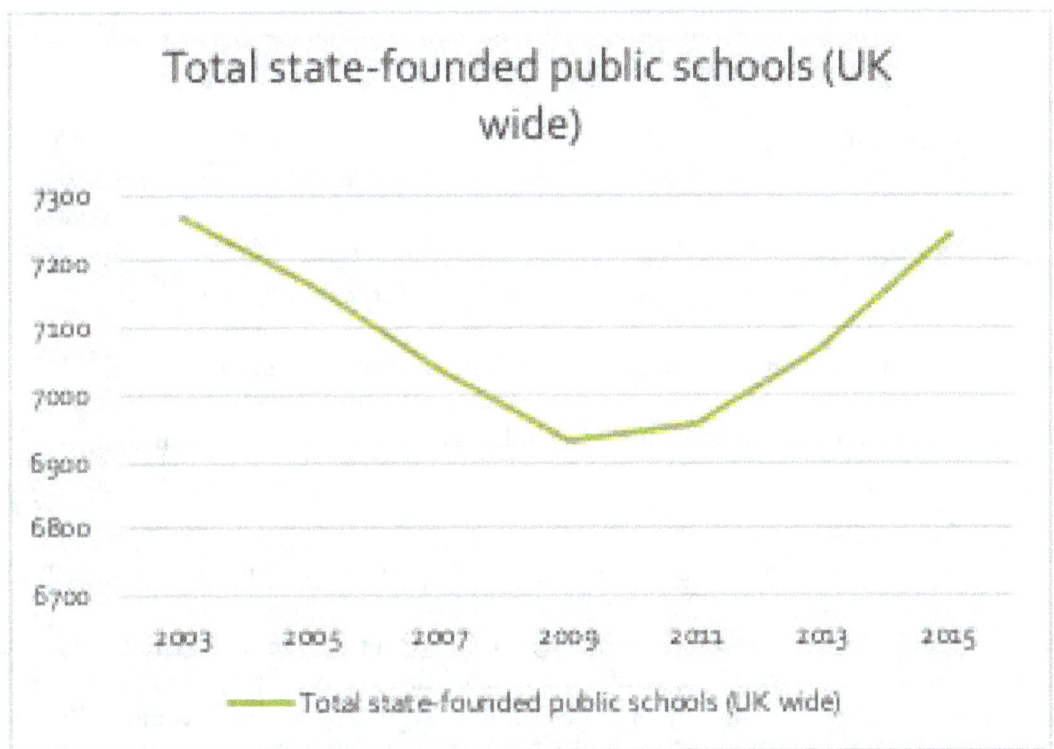

Total state-founded number of public schools (Gov.uk, 2015)

	2003	2005	2007	2009	2011	2013	2015
Nursery/primary schools	4191	4093	4004	3970	4025	4197	4400
Secondary schools	2994	2994	2955	2883	2839	2780	2740
Special schools	83	78	77	77	78	82	87
Alternative provision settings					14	13	12
Total schools	7267	7165	7037	6930	6955	7072	7240

UK state-founded number of public schools (Gov.uk, 2015)

35

The number of pupils is relevant as it influences the amount of exposure which solar heating and cooling can have within the UK if pupils are taught about it in schools.

	2010	2012	2014
<5	811	855	878
5-10	3,501	3,572	3,766
11-15	3,119	3,064	2,976
16+	512	526	544
All ages	7,944	8,018	8,164
Only 5-15	7,431	7,492	7,620

Pupils attending state-founded schools (Gov.uk, 2015)

According to the Office for National Statistics in 2015, from 100 families, 46% had one child, and 40% had two children (Office for National Statistics, 2015).

Number of dependent children	2004	2014
One child	43%	46%
Two children	41%	40%
Three or more children	16%	15%

Office for National Statistics - families and households (Office for National Statistics, 2015)

The two tables above confirm that over 5000 families will send their children to state-funded schools:

Children age	Children in state-founded schools	Number of families
0-16+	8,164	5388
5-15	7,620	5029

Families sending children to public-founded schools (Office for National Statistics, 2015)

When referring to publicly funded schools, given the ratio between the number of families with one child and number of families with two children, it can be estimated that for every ten children who are being taught about solar heating and cooling systems and their importance, 6.6 families would indirectly benefit by acknowledging the information through their child.

Children age	Children in state-founded schools	Number of families
0-16+	8,164	5388
Average	10	6.599

For every 10 pupils taught about SHC, 6.6 families would benefit by acknowledging

This shows the potential with which solar heating and cooling can spread details within schools about ecological and financial benefits if implemented.

Deployment and maintenance costs

Solar heating and cooling systems are affordable for families, but the question is if installing such systems is also accessible for schools. Statistics show that installing such systems for a household will have a return of the investment within a period of 3-6 years depending on the number of household members and the usage of hot water. What's also important to note is that while the number of solar panels increases for a school, compared to a house, the number of pupils will also increase – therefore school revenue will also increase. This process means that installing a considerably larger number of solar panels doesn't necessarily mean that the return on the investment will be delayed.

Considering a standard household with four members from which two work, meaning only 50% of the members can contribute to the repayments. They will install four panels and will obtain a full return on their investment within six years (Association, Solar Energy Industries, 2016). Pupils predominantly use schools, and as a commercial service, it is being covered financially by parents, loans and government schemes. Therefore it can be concluded that 90% of the members who utilise the hot water will have a contribution to the repayments of the investment. The estimation of 90% implies that 10% is the staff.

According to the Department of Education, 2.7 million pupils attend 4722 academies or free schools every year. This information suggests that there is an average number of 571 pupils per school (Drake, 2015). When assessing the average number of panels installed as part of the initiative organised by Oxfordshire County Council and Low Carbon Hub, an estimation of 100 solar panels per school installed purely for solar heating and cooling can be obtained. Several schools decided to set up more than the previous estimation however as their scope was to go beyond just producing hot water and heat within classes, their objective was to produce electricity and transfer it to the national grid.

Members in total	Working members / sources of revenue	Number of panels	Return of investment (years)
4[17]	2[18]	4	6 years
571[19]	571[20]	100	?

Estimation on return on the investment for schools (multiple sources listed above)

This research will consider the amount of time in which the return of the investment can be achieved, subject to ideal weather conditions. The ideal weather conditions applicable to the UK are recorded during June-July months when maximum exposure to the Sun takes place.

Members contributing to ROI[21]	Number of panels	Return of investment (years)
50%	4[22]	6[23] years
100%	100[24]	0.58[25] years

Estimation on an average school's capacity to absorb costs for 100 solar panels

When considering the above information, it can be estimated that such an investment can be absorbed in less than 12 months by a school which has the above characteristics.

---

[17] Two parents and two pupils – standard family format
[18] Two parents working only
[19] Average number of pupils per school
[20] Same number as "members in total" (all pupils represent a source of revenue)
[21] Return of investment
[22] Four panels per household, average identified within the research
[23] Maximum length of time required in order to obtain return on the investment
[24] Estimated number of panels based on all applications included within the research
[25] Estimated return on investment based on previously added factors

Set up

Costs vary based on deployment size, application characteristics, complexity, market characteristics and performance but generally speaking a one-off upfront and frequent maintenance is assumed.
Key elements to consider during the setup stage:

- Solar collectors and similar equipment;
- Mounting components, storage containers and similar devices, filters, plumbing or similar;
- Labouring costs for installation and assembly and scaffolding.

Depending on the market and number of competitors offering installation services, in small set up cases, installation costs can be as high as 50% of the equipment expenses. Large setups generally speaking offer better resource options; therefore, prices are slightly more competitive.

Maintenance

Costs associated with maintenance are low as the technology doesn't require fuel, and they run on minimal electricity usage for ancillary purposes such as water pumps.

Obstruction found on solar panels at Banbury Academy (Solar Panel Cleaners, 2004)

Obstacles can obstruct solar panels from having full exposure to the Sun. Permanent damage can also take place by having the solar panels scratched – this occurs with objects thrown. The majority of solar panels are built to resist elements of nature.

Obstruction at Banbury Academy, damaging solar panels (Solar Panel Cleaners, 2004)

Parts which do require servicing are the absorption and adsorption chillers for which trained experienced technicians will follow a standard cleaning and assessment method statement. Maintenance of solar panels is constituted from regular cleaning to make sure dust or dirt will not obstruct solar radiation from reaching the receptors (Williams, 2014). Furthermore, unremoved obstructions have the potential to scratch panels and produce permanent damage which will then require replacement of parts.

Initiatives assessed for the purpose of the research project

Welsh European Funding Office

Welsh European Founding Office (WEFO) invested funds within SBED ("sustainable building envelope demonstration") project which aims mainly to monitor eight TSC ("transpired solar collector") installations within Wales and England for 24 months, to produce data for further development.

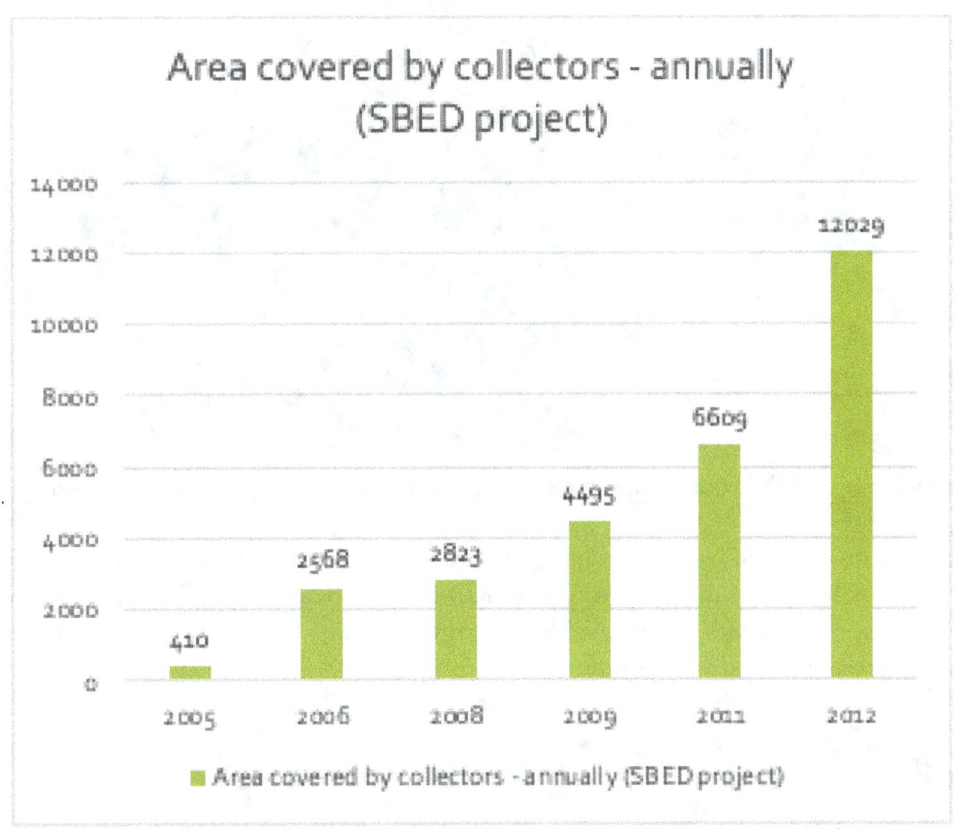

SBED - total area covered by collectors (Williams & Manager, 2016)

The details provided by SBED suggest a rapid growth within a period of only seven years, given that the project started from less than four system deployments in 2005.

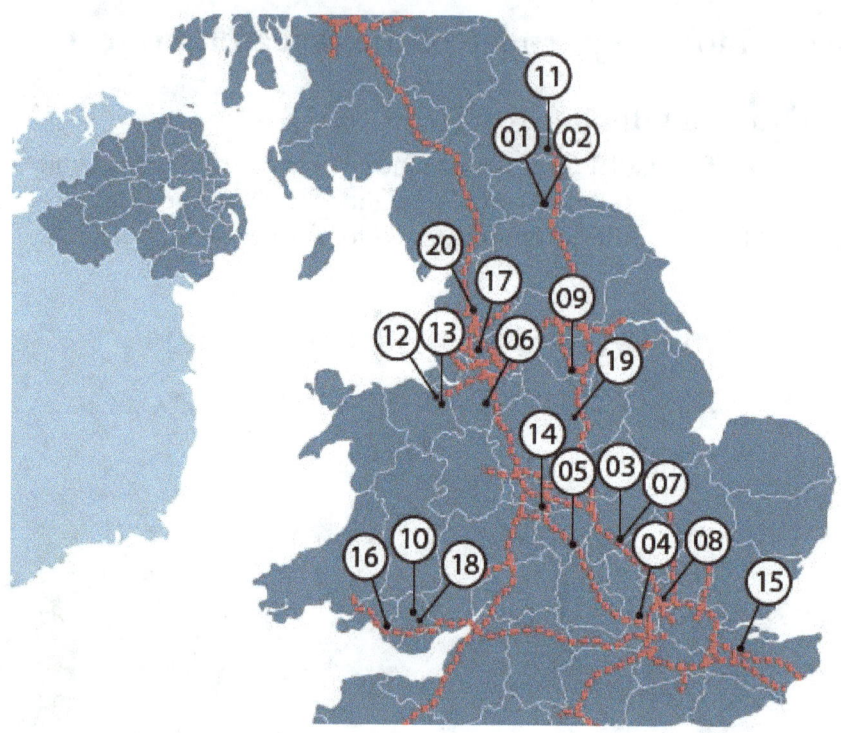

Total number of TSC deployments as of December 2013 (Williams & Manager, 2016)

Transpired solar collector technology was initiated in 2005 in Evenwood, County Durham, UK, it is still being developed, and it hasn't reached maturity, but it aims to preheat using solar radiation the ventilation air supply of buildings. The technology is suitable for large buildings, mainly for commercial purposes, and it has the potential to be deployed within schools.

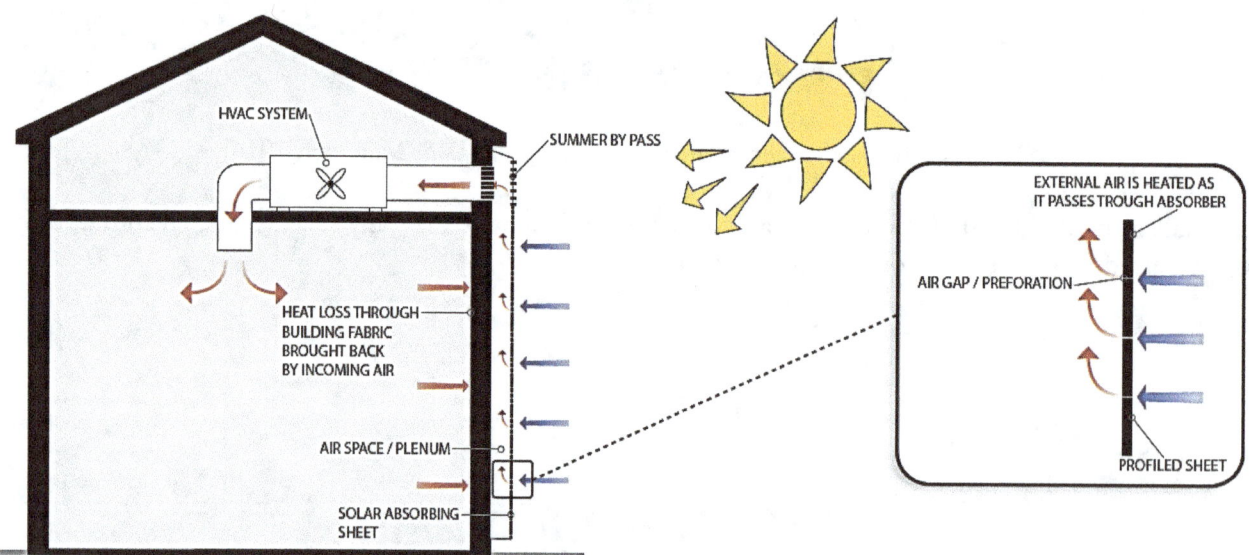

Transpired solar collector system diagram (SBED, 2006)

Given the transpired solar collector technology is still relatively new within the UK, the organisers have defined ideal packages based on suitability for various types of deployments, in terms of energy produced and installation costs encountered.

System deployment type	Expected TSC[26] area
Residential	40m²
Offices	200m²
Industrial	100m²
Schools	150m²

TSC recommended exposure areas for solar panels (Shanks, 2002)

Largest solar park in Scotland

British Solar Renewables and BWE Partnership (Shanks, 2002) have announced progress on the New Mains of Guynd £10 million investment solar park, expected to enter production by June 2016.

New Mains of Guynd Scotland's largest solar park (British Solar Renewables Limited, 2016)

The 9.5MW project is an example of the extent to which solar radiation can be used, as the scope is to use invertors and feedback to national grid electricity, produced entirely from Sun radiation (British Solar Renewables Limited, 2016).

Oxfordshire Council and Low Carbon Hub

Oxfordshire County Council in collaboration with Oxford City Council and Low Carbon Hub Organization have engaged with local schools within the borough to help them improve their energy efficiency, by providing them with access to specific loans or energy surveys to set up solar panels and install Solar Heating and Cooling systems over their infrastructures.

The project aims to implement solar heating and cooling and to add 5000 solar panels set up across 23 schools mentoring 9000 pupils, within Oxfordshire by 3rd quarter of 2016, as part of project OxFutures. The essential requirement for each school was for it to benefit from at least 20m2 exposure to Sun to position the solar panels (Hub, 2015). Oxfordshire County Council has secured £15 million in funding from Intelligent Energy Europe and has tendered the OxFutures project to principal contractor Carillion.

Institution name	Pupils	Solar panels
Banbury Academy	1085	194
Orchard Fields School	420	384
Bure Park Primary	520	240
Charlbury Primary School	210	96
Cheney School	1400	400
Eynsham Village Hall	410	16
Larkrise Primary School	455	78
Middle Barton Primary	157	40
Nettlebed Community School	105	115
The Warriner School	1115	400

---

[26] Transpired solar collector

| Wheatley Park School | 1005 | 195 |

Oxfordshire Council and Low Carbon Hub Organization integration of schools (Hub, 2015)

Various enquiries were made towards the Department for Education (Education.gov.uk, 2010) to extract details about the volume of pupils attending school. Other sources of information used were to obtain information on the number of panels used for each school (Herring, 2015), (Rowell & Simpson, 2016).

Largest solar panel installation in a primary school (Renewable Energy Focus, 2009)

As part of this project, main stakeholder Oxfordshire Country Council promises to provide free energy surveys to schools, automatic meter technology and useful loans to equip as many schools as possible with technology which aims to save energy and decrease pollution (CAG Oxfordshire Community Action Groups, 2015).

	2010-2011	2011-2012	2012-2013	2013-2014	2014-2015
Reduce usage of gas, kerosene, oil, LPG, petrol, diesel	25543	17948	22293	19365	10873
Reduce purchased electricity	35358	31865	33264	31100	25228
Energy used for contractor's offices/business travel/energy transmission	7140	6015	5894	6567	5588
Total	68041	55828	61451	57032	41689

Oxfordshire Council Green Gas Report - (CAG Oxfordshire Community Action Groups, 2015)

The Oxfordshire Council driven project was designed beyond the simple purpose of solar heating and cooling. Apart from producing hot water for use, several schools applied to be equipped with technology in order to produce electricity from solar radiation, electricity which was then used or returned to the national grid. The above statistics reflect to the overall performance of the council's project and it shows the extent to which solar heating and cooling can be deployed in order to maximise its efficiency.

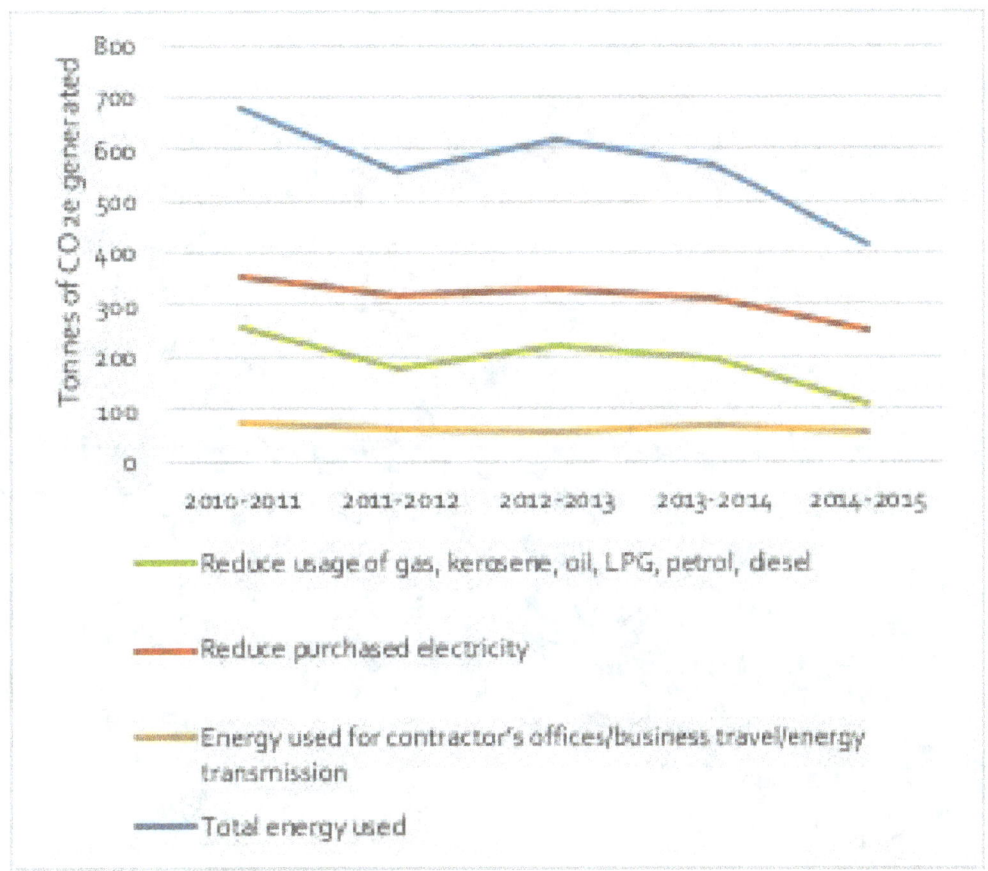

Pollution reduction in Oxfordshire (CAG Oxfordshire Community Action Groups, 2015)

Statements released from Oxfordshire Council shows the decrease within the annually published CO2e, as a direct result of the OxFutures project.

Mexborough High School, South Yorkshire

The Mexborough High School is located in Doncaster and mentors 812 students between the ages of 11 and 18 years old.

The school management represented by N G Bailey contracted Norwest Holst, to integrate a ground surface heat pump into a new thermal water heating system using solar heating and cooling principles, in an attempt to decrease electricity costs and to minimise the negative impact on the ecosystem.

Various 3rd parties were involved indirectly – all supporting solar heating and cooling principles: Carbon Trust Group, Mitsubishi Domestic Heating Solutions and smaller groups representing eco-friendly initiatives.

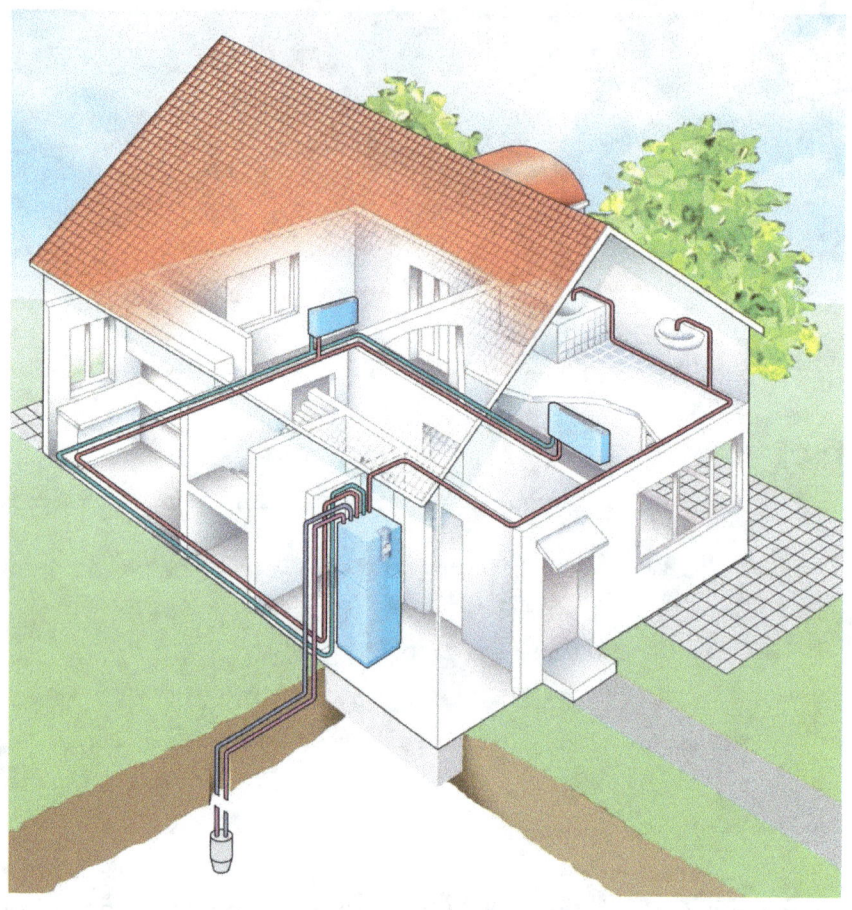

General solar thermal delivery mechanism diagram (Solarta, 2004)

The above image shows the distinct components installed within the Mexborough High School project, except within a simplified Revit drawing. The project amounted from the installation of the solar collectors on a roof section, using copper pipework to transport heat using antifreeze liquid, a main thermally isolated cylinder and connections to the existing pipework and radiators.

Stiebel Eltron ground source heat pump (Revolution Power, 2004)

Mexborough High School solar heating and the cooling project was designed to deliver both hot water and room heating following the legal standards, and it was estimated that during the summer the boiler would not have been required after the project upgrade. The heat pump was designed to deliver water at 60°C, and it had a COP[27] of 4.5 at ~35°C. It was also estimated that the overall contribution of the solar heating and cooling system should have produced 50-70% from the total volume of hot water used for heating per annum (Revolution Power Solar Thermal Heating, 2006). The pumped/forced circulation system upgrade was considered a success by all parties involved, including Carbon Trust Group.

Edlington High School, South Yorkshire

Home to 273 pupils, Edlington Victoria School's heating system was upgraded by Norwest Holst contractors, following coordination of Revolution Power Solar Thermal Heating Limited. The upgrade was constituted from a 500-litre cylinder connected to a solar heating system fitted on the building's façade, a pump station and pipework to join the new heating system to existing infrastructure.

Schuco solar thermal system (Revolution Power Solar Thermal Heating Limited, 2005)

With a 95% solar radiation absorption, the above panels provided approximately 450 litres of hot water per day during summer (Revolution Power Solar Thermal Heating Limited, 2005).

---

[27] Coefficient of performance

Pump and 500 litres cylinder (Revolution Power Solar Thermal Heating Limited, 2005)

Eynsham Pavilion

Following Carbon Trust Group's initiative, Eynsham Pavilion School in Oxfordshire has been upgraded with a solar heating and cooling system using 16 solar panels which should produce 40% savings from the energy consumed for heating purposes.

Figure 1 - Eynsham Pavilion solar panels (Solar Century, 2015)

The Parish Council objective was to reduce both energy bills and carbon footprint. Home to 406 pupils, the management decided to use a solar heating technique primarily for heating purposes but also during the summer periods they installed inverters to produce electricity and decrease the amount of consumption from the national grid (Solar Century, 2015). Gordon Beach, Chairman of Eynsham Parish Council, emphasized the ease with which the solar panels were installed, within a period of one day, yet they will serve for the next 25 years. He also added that the performance of the solar panels is outstanding as they can capture radiation even through cloudy days.

Larkrise Primary School

During the summer of 2014, Larkrise Primary School was integrated within the Low Carbon Hub, having installed 78 solar panels on its roof during the summer holidays.

Larkrise Primary School - solar heating and cooling project (Oxford Mail, 2015)

The school's management has estimated that the panels save 25% from the energy which would have usually been required for heating purposes every year and as with the other projects, the solar heating and cooling system should suffice the hot water supply over the summer (Oxford Mail, 2015).

Solar-combi - heat storage designed to fit solar heating and cooling

Solar-combi was designed in 2007 to put together two components which were previously already connected; however, never initially created for each other: solar heating technology with temperature storage technology. The aim is to eliminate loss by introducing innovation within both technologies and by delivering a complete package, ready to be deployed.

The project was aiming to satisfy demands usually required by schools and large institutions (Sources, Centre for Renewable Energy, 2009). The project was tested in various favourable European countries including Greece, Italy, Spain and Austria and the outcome consisted of more effective thermal isolation used for storage tanks so that the duration with which hot water would have remained available, would have been longer. It needs to be noted that the countries where the Solar-combi project took place to benefit from above-the-average temperatures during the warm season and therefore the outcomes of the experiments carried out as part of the project are not necessarily applicable for other countries benefiting from a different climate such as the United Kingdom or northern countries.

Abbots Bromley School for Girls

A new solar panel system replaced the already existing 30 years old one. The new system was designed with 32 solar panels in 2 sections, 16 panels each. The maximum capacity for a field is 50 panels. Buderus solar panels are designed for large scale deployments, i.e. industrial scales.

Integration with specialised boiler features will achieve even less consumption as boiler start-up times can be decreased to when the demand requires water to be heated – instead of chronologically controlled boiler engagement. This setup reduces electricity by 24% - by avoiding unnecessary boiler start-up procedures (Buderus, 2016). The integration goes as far as fitting a specific mounting system SKS, 4th generation, which makes easy the repositioning of the flat place collectors. The critical system behind the panels consists of a continuous copper pipe used for fluid circulation by distributing the temperature across the tubing.

Results and analysis

TSC - suitable for industrial purposes

Despite being new, some technologies such as TSC (Transpired solar collectors) have spread relatively fast and are now covering 12500m2 of land with only 20 system deployments. Provided this solution is suitable mainly for industrial purposes, it can be concluded that each deployment has a considerable area covered for installing solar panels. The Sustainable Building Envelope Demonstration project aims to assess the ongoing efficiency of TSC and to monitor maintenance needs and the volume of new deployments. It is worth mentioning that industrial dwellings such as halls or hangars have a considerable amount of area exposed to the Sun and the need to have side windows is not as crucial as it would be within schools; therefore it can be concluded that TSC is suitable for industrial purposes, despite it being opposite to school and academic institutions.

Classic aircraft hangar (SCG GRP, 2004)

Increase awareness through campaigns within schools

The research paper proved that over 5000 families are sending their children every year to UK public-funded schools. If all public schools would agree to implement solar heating and cooling programs which can include both system implementation and awareness campaigns, 5000 families will be made aware of the importance of running solar heating and cooling, within the UK. This opportunity clearly shows the potential of solar heating and cooling if adequate awareness would be created and sustained by family members of pupils.

Variety of consumptions suitable for any school

Opposite to TSC, classic solar heating and cooling systems can be used for small schools or areas from within schools where the consumption is less intense. Service providers can set up portable systems which can be deployed at the beginning of the season and removed when not required for ease of access.

Portable SHC system – roof of dwelling (Earthnet Energy, 2006)

Such systems do not require civil or construction-related works to take place and can be achieved using skilled labour, presumably already engaged with regular school maintenance. More powerful systems are available, which are also suitable for schools and premises used for academic purposes. The compound parabolic concentrator is designed for a concentration ratio of <2 and works well with non-tracking systems because it can diffuse radiation. Maximum reflector collectors can achieve high efficiency in spring and autumn, though in the summer it doesn't heat due to an asymmetric reflector.

Example of portable concentrated solar technologies system (Inhabitat, 2004)

For requiring a high amount of heat – schools that have a large number of classes to heat, concentrating solar technologies can be used if positioned on top of a roof. It may not work with UK weather, but if optimized, i.e. track both Y and X it can work. These systems can also be used for industrial purposes.

Structural analysis - vertical positioning - schools - cover walls

Glazed air collectors (Green Building Advisor, 2003)

Between glazed and unglazed air collectors, the glazed collectors are suitable for space/building air heating. Such collectors are fervently positioned on the outside wall of a building, due to its exposure to the Sun.

Each council should run its own campaign

Following the example of Oxfordshire County Council and Oxfordshire Council, all councils should consider plans to create awareness campaigns and promote energy-saving mechanisms within the schools administrated within their borders following programs such as OxFutures, funded by Intelligent Energy Europe (European Comission, 2015). As shown by the research paper, Oxfordshire County Council have managed to reduce the amount of pollution by 39% over a period of 5 years, only by investing within local schools.

The effects of the above project could be more significant if other public institutions will also benefit from solar heating and cooling technology such as universities, hospitals and police stations. It is essential to take note of the fact that Oxfordshire County Council considers Oxfordshire to be a relatively rural county (Council, 2012). This element is vital because rural areas benefit from lower costs for access to land, which makes a project affordable as it will utilize a considerable amount of land to install solar panels.

At initial assessment, the reader could say that solar heating and cooling is feasible only within rural areas in an attempt to undermine its potential, but having a closer look at England's, Wales', Scotland's and Northern Ireland's local authority classification (Pateman, 2015), it becomes evident that the majority local authorities benefit from enough rural areas to deploy the eco-friendly technology within – solar heating and cooling deployment can take place.

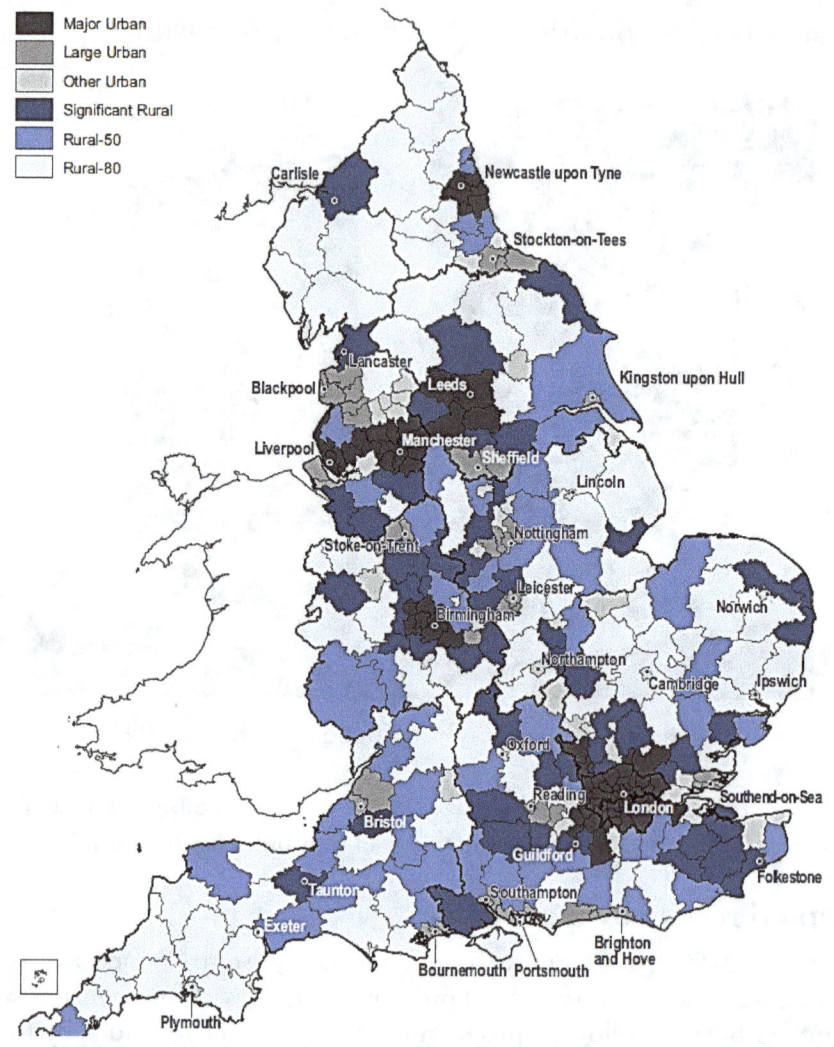

England local authority rural/urban classification (Pateman, 2015)

An exciting observation refers to the overcrowded urban areas such as the City of London and the possibility to implement solar heating and to cool within such environments. From the technical perspective as long as the structure allows positioning solar panels and as long as there is space to store isolated tanks to maintain the temperature of the hot water, solar heating and cooling can be implemented even within busy urban areas. There are seven schools within the M25 Greater London region who have installed solar heating systems within their infrastructure (Solar Schools, 2010).

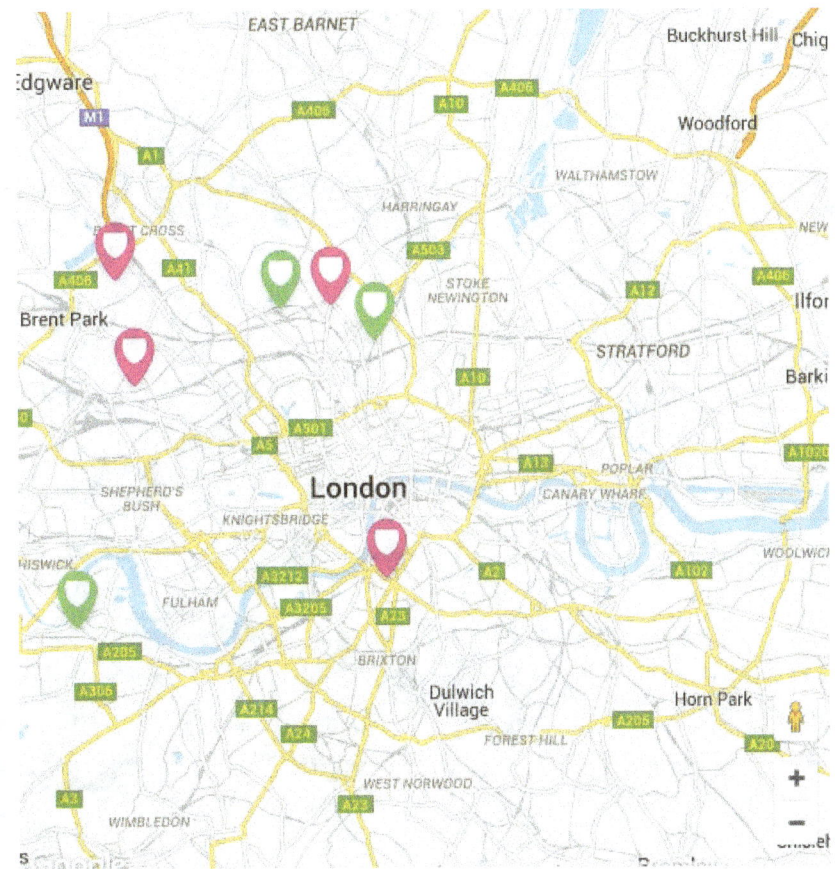

Schools within London which use solar heating and cooling (Solar Schools, 2010)

The fact that only seven schools from within Greater London have installed a solar heating and cooling systems can be caused by a lack of exposure to the Sun, potentially due to obstructions such as tall buildings. Another reason can be consisted of high competitiveness, disallowing such long term investments to take place.

Another potential reason can be the lack of awareness about solar heating and cooling or the fact that the general opinion from the market discourages such applications as a result of the British-specific weather. An essential action to take to overcome this is to create a digital string campaign for informing parents and schools about the benefits of implementing solar heating and cooling, specifying both environmental and financial impacts.

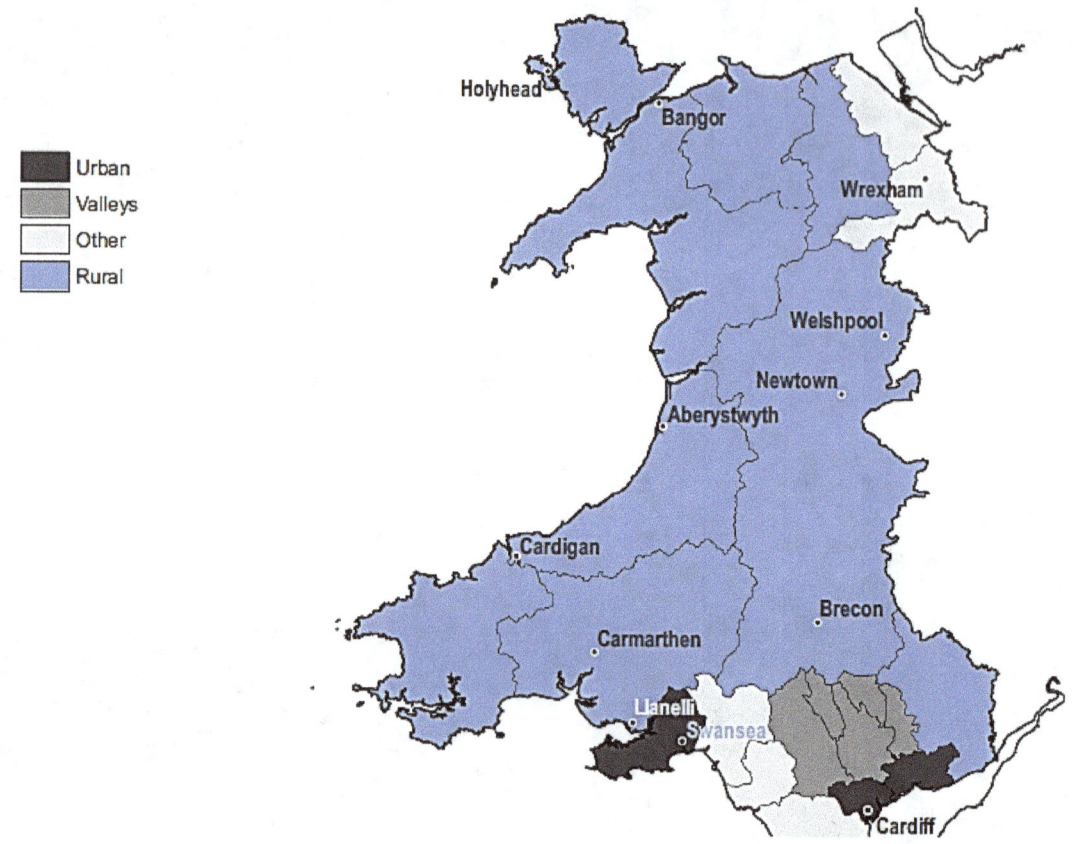

Wales local authority rural/urban classification (Pateman, 2015)

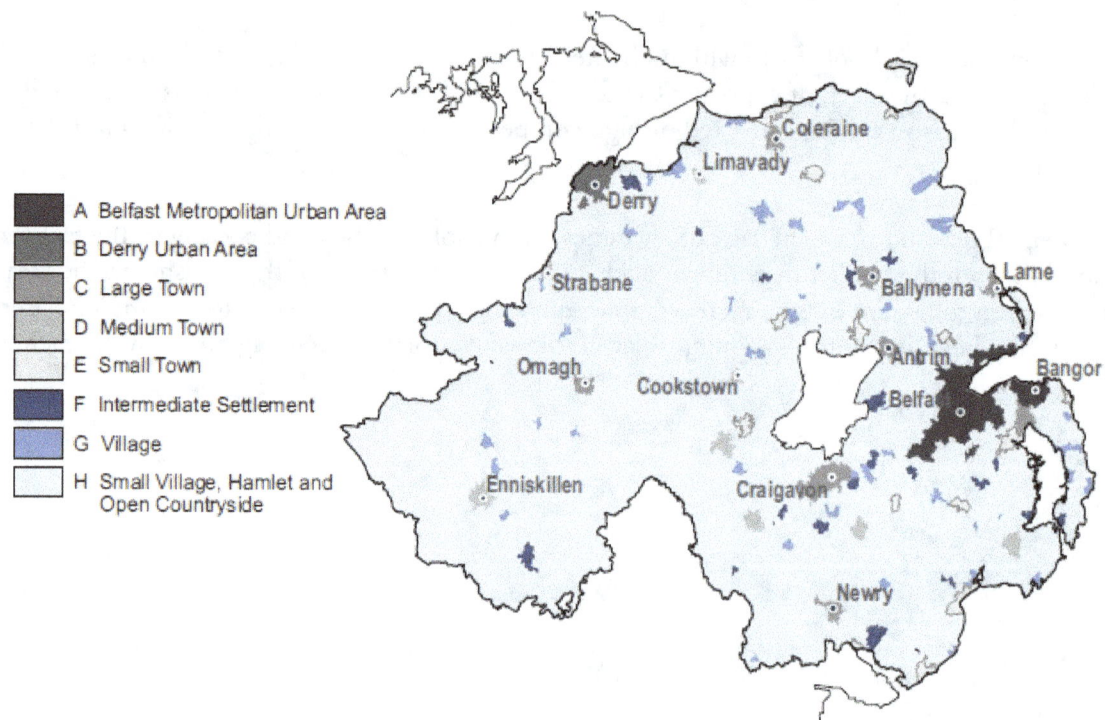

Northern Ireland local authority rural/urban classification (Pateman, 2015)

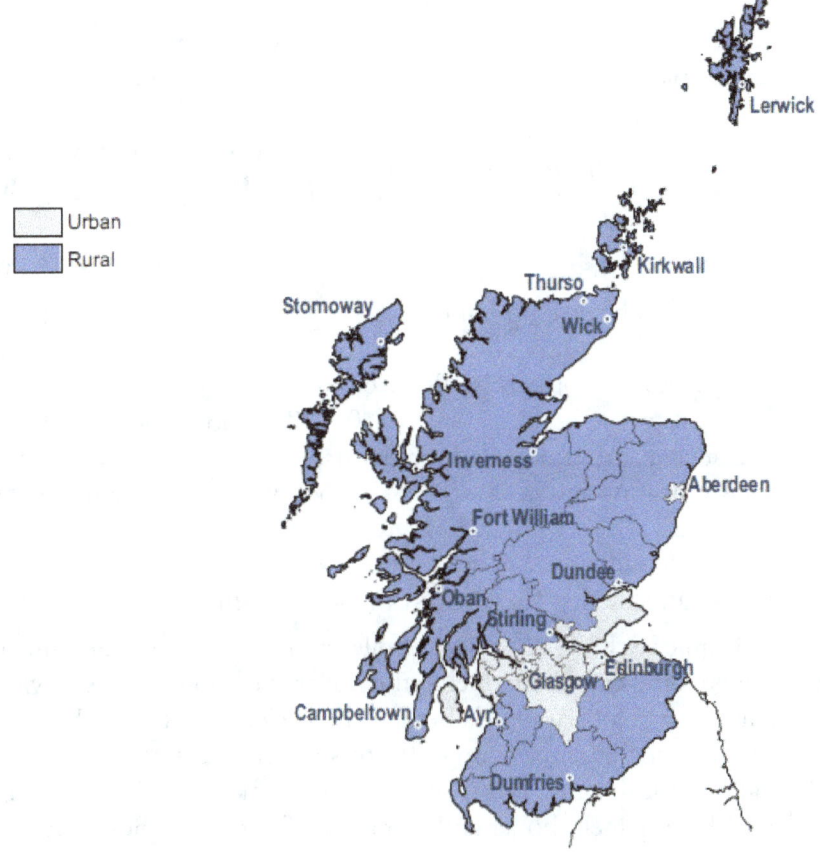

Scotland local authority rural/urban classification (Pateman, 2015)

Effects of a coordinated campaign within a local administration

As demonstrated above Oxfordshire County Council has engaged with Low Carbon Hub within Oxfordshire to set up the Solar Schools Scheme, aiming to install 5000 solar panels within 23 schools as part of phase one (2014-2015). The principal objective of the initiative is to reduce electricity costs and carbon footprint by using solar radiation for heating purposes and secondly for producing clean energy.

Eleven schools are considered above, located within Oxfordshire, integrated within the Low Carbon Hub program, and are mentoring a total number of 6882 pupils. These eleven schools have installed a total number of 2158 solar panels. This proves the popularity that Low Carbon Hub has within the schools located in Oxfordshire and the ease with which such an initiative would grow if supported by a local authority.

The above figures also show the extent to which solar panels would be used if a robust program was in place, a few of the schools integrated within Low Carbon Hub have chosen to acquire solar panels not only for solar heating and cooling but also to produce energy using invertors and to sell it back to the national grid. The difference can be noticed when referring to the ratio between the number of students mentored and the number of solar panels installed:

Institution name	Number of pupils	Number of solar panels
Eynsham Village Hall	410	16
Nettlebed Community School	105	115

Various ratios between number of pupils and number of solar panels

Schools with a lower number of solar panels would use them only for water heating purposes while schools that install a higher number of panels would look towards producing electricity.

Institution name	Number of pupils	Number of solar panels	Solar panels / Pupils (%)
Eynsham Village Hall	410	16	4%
Nettlebed Community School	105	115	110%

Example of ratios between number of solar panels and number of pupils

The analysis conducted on the eleven subject schools revealed the ratios between the number of solar panels over the number of pupils varies between 4% to 110%. The above shows that the scalability with which solar panels can be installed and deployed. At initial stages, it can be connected to pipework and storage tanks and can facilitate heating and hot water for school premises. If required at advanced stages, it can also be connected to inverters which will produce green electricity.

Compact package decreases long term costs and increases efficiency

As shown within Abbots Bromley School, acquiring a boiler manufactured by the same company that builds the panels decreases ongoing costs by reducing the boiler start-up times, as it will be programmed to turn on the heating system only when solar panels become inefficient due to weather changes or end of daylight. The integration list also includes mounting supports which facilitate relocation or the possibility to change the inclination angle with little or no effort through electronic means. The benefits of using a compacted system which includes panels, boiler and supports, within UK schools will overcome the weather disadvantages by maximising the efficiency for heat absorption.

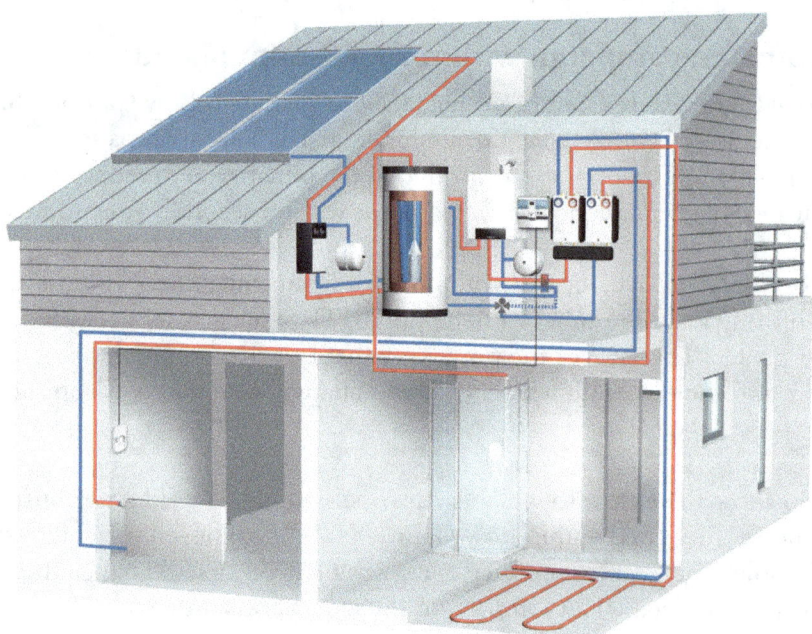

Buderus integration and deployment capabilities (Baulinks, 2010)

Evacuated tube collectors

Evacuated tube collectors represent a better technology for cloudy environments (Matters, 2014) compared to flat plate collectors, because of their increased efficiency up to 163%. Such collectors have better exposure to radiation from the Sun and can even work within negative temperatures. In terms of efficiency, evacuated tube collectors require a smaller area than flat plate collectors to achieve the same results. In terms of maintenance, statistics show that flat plate collectors develop corrosion more often than evacuated tube collectors. For the research paper, evacuated tube collectors prove to be a more suitable solution for UK schools given the often cloudy weather which can obstruct solar radiation.

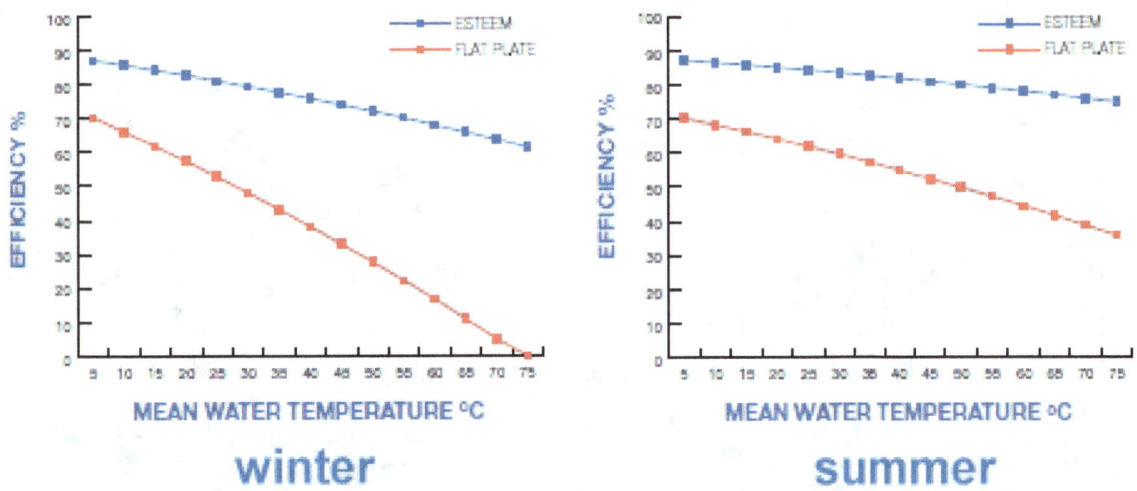

Evacuated vs flat tube collectors - winter vs summer (Matters, 2014)

This information suggests that evacuated tube collectors would be suitable for UK schools given that UK weather often produces clouds.

Flat plate collectors - do not require Sun tracking

Given that flat plate collectors can absorb the energy coming from all directions above absorbers, mechanical methods to track the Sun are not required. Therefore, for setups which require over ten panels, which would be suitable for UK schools, it would be easier to use flat plate collectors and not have to perform angle adjustments several times per day to increase efficiency. Fixing the collector panels to metallic supports will increase stability in windy environments. Therefore, for rural areas where wind circulation is higher than within urban areas, flat plate collectors may be suitable.

Unglazed flat plate collectors - for low temperatures with low or no wind

As described above, unglazed flat plate collectors would be ideal for UK schools given average outside temperatures can settle around 30°C. For not having any glazing the plates capture energy from the Sun even when outside temperatures are less than 25°C. For not having a layer of isolation, the plates can lose heat if the environment is windy; therefore, the installer must assure the environment where plates would be installed is not windy.

Storage heat – feasible for schools

As shown above, the use of storage for heat is fundamentally essential to extend the delivery period for hot water to up to 24 hours. Schools benefit for storage space as they are standalone dwellings with multiple rooms available for use. Dedicating one room for storing isolated tanks will extend the period for when hot water is being delivered to cover through the night. Storage for hot water involves a financial impact; however, it increases efficiency for providing hot water outside the intervals where there is exposure to the Sun, i.e. during the night time or early hours of the morning.

Glazed air collectors, economical, no need to tilt

Another economical and efficient solution for UK schools consists from glazed air collectors who do not require tilting, can be installed with low costs and can be attached to walls exposed to Sun and used as long term solar heating systems.

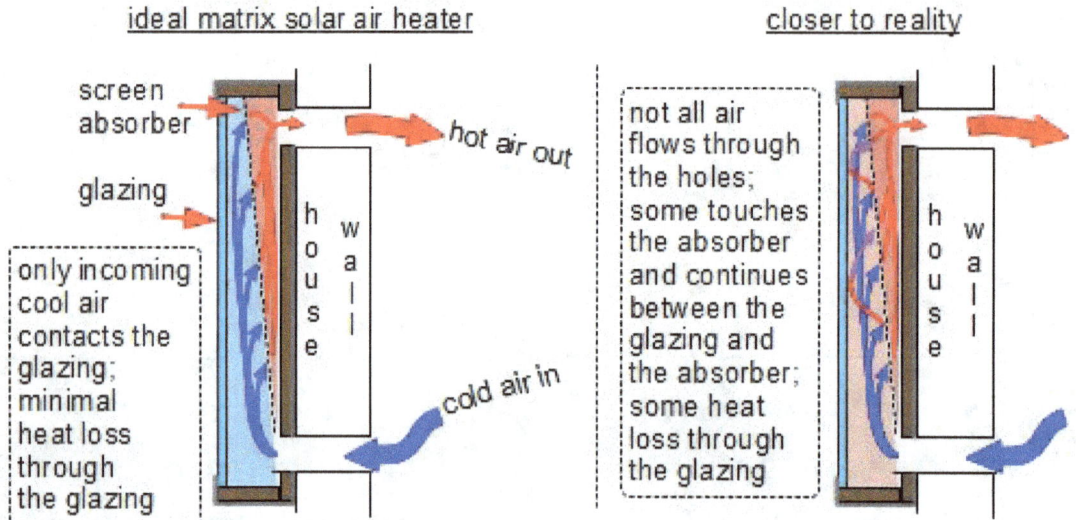

Figure 2 - Glazed air collectors – diagram (Rimstar.org, 2016)

Another reason for which glazed air collectors would be suitable for UK schools is the fact that once such investments are made, it is expected to cover as many sides of the building as required to obtain maximum efficiency. For usually being independent dwellings, schools offer all four walls exposed to the elements, available to have glazed air collectors position across the ones positioned relevant to where solar radiation would reach solar panels as Earth revolves around the Sun.

Discussion and conclusions

Solar heating and cooling is a mature technology which provides multiple methods of implementation suitable for various types of deployments, all appropriate for UK weather conditions. The extent to which solar heating and cooling are being implemented allows solar panels to be deployed to capture energy from the Sun, not only for domestic heating purposes but also to generate electricity and sell it to the national grid.

The research paper compared all points discussed above and positioned them within the context of being used within UK schools. The outcome from the research suggests that UK schools would be an ideal environment for solar heating and cooling integration for a variety of reasons presented above, including the amount of space available to position solar panels, the potential savings and the positive impact which solar heating and cooling will have on the environment, amongst others. Before deploying such systems, every school needs to assess the consumption of hot water and energy used for heating air. This evaluation is required to identify suitable applications for deployment.

Local administrative support

Local administrative initiatives support solar heating and cooling in various parts of the UK and deployments have already taken place with 4-16 solar panels solely to generate hot water and up to 400 solar panels for advanced purposes of producing pure electricity from energy captured from the Sun.

The extent to which Oxfordshire Council run their campaign involves a total number of 5000 panels to be installed within 23+ local schools. The benefits of having an organised action consist in the ease with which schools can subscribe to such programs, the documentation and procedures which are applied uniformly to all school in the partnership, the financial and ecological impacts felt by reducing the energy required for heat generation.

The paper looked at the rural/urban distribution charts and recommended for all counties to carry out internal projects and motivate schools to subscribe to such initiatives. The paper also acknowledges the impact of educating pupils and creating awareness campaigns about the importance of solar heating and cooling techniques. One of the effects of such local administrative driven projects is to make parents aware of the economic and environmental benefits of using solar panels and potentially trigger interest for such integrations for domestic purposes.

The variety of solar heating and cooling systems

Various branches of solar heating and cooling become more popular such as TSC, evacuated tubes, circulation systems or flat plates. When deploying such systems, the aim is to identify the most suitable technology considering a variety of factors.

When looking at schools, specific questions are being asked about the environment, such as autonomy, legal requirements in terms of class temperature. One of the aspects which will influence the project specifications refers to the volume of hot water required to be used on average per day. Smallest implementations of solar heating and cooling use a single tank of 500 litres which can be used at any moment in time. Structural details about the dwelling are also relevant as solar panels will need to be positioned on one of the four walls and over the roof. Schools with one floor at least would benefit from the possibility to place solar panels on walls at a safe distance from pupils or members of staff.

A coordinated centralised initial will also influence the budget allocated for such an implementation as different schools benefit from a variety of budgets, yet a loan or a structured scheme will support schools' financial capacity. It will also enable them to subscribe for most effective solutions.

Position concerning the exposure to Sun

For referring to schools which are generally independent buildings and have four walls exposed to the elements along with a roof, it is reasonable to presume that there is enough space to position solar panels particularly if positioned vertically on walls.

Tilting is facilitated even on the side of the walls using supports which allow restrictive rotation for solar panels to capture more solar energy as Earth rotates around the Sun. Positioning the solar panels on the roof will provide a more extensive area making solar panel tilting or rotation more straightforward, a potentially cheaper option. Panels positioned on the side of a building will not provide as much efficiency as panels placed on the roof, because of the limited amount of exposure.

Flat plate collectors

For economical deployments, a technical advantage can assist in eliminating the need to track Sun movement by installing flat plate collectors which can absorb heat even if not directed perpendicularly to the panel's surface. This option becomes extremely efficient when deploying structures with a considerable number of solar panels up to 400, as shown within the research paper when installing tilting mechanisms for such as a large number of solar panels would involve considerably more significant investment.

Extending the storage capacity

In terms of storage, the smallest tank identified within the research project holds a volume of 500 litres of hot water. A simple and effective way to extend the amount of hot water stored is to use larger capacities of storage tanks. This method depends on the isolation technique as the water stored must remain hot. For large volumes of usage, multiple storage tanks could be provided and positioned of such that will cover different building areas.

Compact solutions bring benefits

This research paper also demonstrates the importance of identifying a compact solution which will provide a boiler, solar panels and potentially mountable racks that can track and orientate the panels for maximum effectiveness, all being manufactured by the same company or fully compatible to operate together. An adequate system will suspend the boiler's functionality if the tank already contains enough hot water to satisfy demand. This condition will extend the autonomy of the boiler, which will reduce energy consumption and effectively will have a positive impact on the environment.

UK climate and exposure to Sun

The UK climate specifics can be addressed by using unglazed solar panels which work even within temperatures of less than 30°C; however, they must be protected against the wind as not having an isolative layer on top, means that heat entering the absorption plates can easily be dispersed by wind. Unglazed solar panels also work in cloudy environments, making them ideal for UK climates. Adding a protective layer of glazing will increase the operating temperatures making the receptors slightly less receipting under low temperatures but will engage protection against wind. At this stage, it becomes essential to assess the location of the school and the historical weather and wind activity within the area.

In terms of solar cooling, provided the average summer temperatures do not reach the same temperatures encountered within continental Europe within countries such as Spain or Italy, it can be presumed that the cooling aspect is not as important as the solar heating is.

Conclusion

Solar heating and cooling offer multiple options, technologies and resources. It has reached a stage at which it needs engagement from central administration to trigger interest for long term investments within UK schools.

The technology is suitable for UK weather and given it is being addressed to schools; it can benefit from a considerable amount of space for deployment, using multiple techniques to produce both hot water and hot air. The concept of solar heating and cooling is fully scalable up to the point where a school can produce hot water for use from using two solar panels, or where the school can generate electricity from the Sun's energy captured with 400 solar panels. Since not all the electricity produced can be used, it can be sold to the national grid, and by doing so, the school is looking at a rapid return on the investment.

The scalability is also reflected in the type of systems which are available to be installed. As shown in the research paper, there are multiple service providers able to assist with a variety of jobs form installing solar heating and cooling systems up to maintenance and cleaning. Any client will have options to choose from in terms of service providers to use. Solar heating and cooling have all the potential to reduce 50% of the amount of energy consumed for hot water and heat, reducing expenses and minimising the greenhouse effects, which are threatening our climate.

References

1. Alternative Energy, 2004. Alternative energy tutorials (image). [Online] Available at: http://www.alternative-energy-tutorials.com [Accessed 02 02 2016].

2. American Society for Heating, Refrigeration, and Air -Conditioning Engineering, 1977. Methods of Testing to Determine the thermal Performance of Solar Collectors. New York: ASHRAE.

3. Apricus, 2003. Apricus.com.au (Image). [Online] Available at: http://www.apricus.com.au [Accessed 1 01 2016].

4. Aprocus, 2010. Apricus - Evacuated Tubes. [Online] Available at: http://www.apricus.com [Accessed 21 01 2016].

5. Association, Solar Energy Industries, 2016. Solar Heating & Cooling, Washington: Solar Heating & Cooling.

6. Barrow, M., 2016. Woodlands Junior School. [Online] Available at: http://resources.woodlands-junior.kent.sch.uk

7. BASC, 2004. Basc.pnnl.gov (Image). [Online] Available at: http://basc.pnnl.gov [Accessed 01 01 2016].

8. Baulinks, 2010. Baulinks (Image). [Online] Available at: http://www.baulinks.de [Accessed 02 02 2016].

9. Boardman, B., 2007. Home truths: a low-carbon strategy to reduce uk housing emissions by 80% by 2050. Oxford: University of Oxford.

10. British Solar Renewables Limited, 2016. British Solar Renewables (Image). [Online] Available at: http://www.britishsolarrenewables.com [Accessed 5 12 2015].

11. Brown, C., Perisoglou, E., Hall, R. & Stevenson, V., 2013. Transpired Solar Collector Installations in Wales and England. Freiburg, Welsh School of Architecture, Cardiff University.

12. Buderus, 2016. Abbots Bromley School. [Online] Available at: http://www.buderus.co.uk [Accessed 20 2 2016].

13. Build it Solar, 2005. Build it Solar (Image). [Online] Available at: http://www.builditsolar.com [Accessed 01 04 2016].

14. Build it solar, 2010. Build it solar - Experiment. [Online] Available at: http://www.builditsolar.com/ [Accessed 12 02 2016].

15. Building, 2007. Building. [Online] Available at: http://www.building.co.uk [Accessed 19 01 2016].

16. Builditsolar, 2016. Glazing solar collectors - Experiment (Image). [Online] Available at: http://www.builditsolar.com [Accessed 6 4 2016].

17. CAG Oxfordshire Community Action Groups, 2015. Low Carbon Communities Handbook. Oxford, Glance Image.

18. Cali, A. et al., 1999. Low cost, high performance solar air-heating systems using perforated absorbers. Washington: IEA1999.

19. Carbon Trust, 2004. Energy Saving Fact Sheet | Schools, London: Queen's Printer and Controller.

20. Casamajor, A. B. & Parsons, R. E., 1979. Design Guide for Shallow Solar Ponds, New York: Lawrence Livermore Labs Report.

21. CC, M., 2004. Field study and modelling of an Unglazed Transpired Solar Collector System, Raleigh: North Carolina State University.

22. Clean Energy Resource Teams, 2012. Clean Energy Resource Teams (Image). [Online] Available at: http://www.cleanenergyresourceteams.org [Accessed 2 12 2015].

23. Clean Green Energy Zone, 2005. Solar heating and cooling for pools (Image). [Online] Available at: http://www.cleangreenenergyzone.com [Accessed 2 2 2016].

24. Commission, E., 2006. Control of the energy supply and environmental issues. Bruxelles, European Commission Archives.

25. Cordeau, S. & Barrington, S., 2011. Performance of unglazed solar ventilation air pre-heaters for broiler barns. s.l.: Solar Energy.

26. Council, O. C., 2012. Oxfordshire Local Transport Plan, Oxford: Oxfordshire County Council.

27. Darling, D., 2000. The Worlds of David Darling: Amazon.

28. Department for Education and Employment, 1999. The Education (School Premises) Regulations 1999. London: Gov.uk.

29. DfES Pupil Health and Safety Team, 1998. Health and Safety - Responsibility and Powers. London: Gov.uk.

30. Dickenson, W. C., Clark, A. F., Day, J. A. & Wouters, L. F., 1976. The Shallow Solar Pond Energy Conversion System. 1976 ed. New York: Solar Energy.

31. Djebbar, D. R., 2012. Survey of Active Solar Thermal Collectors, Ottawa: Clear Sky Advisors Inc.

32. Drake, R., 2015. Schools, pupils and their characteristics: January 2015, London: Department of Education.

33. Duffie, A. & Beckman, A., 1980. Solar Engineering of Thermal Processes. 1 ed. New York: John Wiley, Sons.

34. Earthnet Energy, 2006. Earthnet Energy (Image). [Online] Available at: http://www.earthnetenergy.com [Accessed 05 01 2016].

35. Ecoactive, 2013. Ecoactive (Image). [Online] Available at: http://www.ecoactive.com.au [Accessed 09 01 2016].

36. Edmound, 2010. Compound Parabolic Concentrators (CPCs), London: Edmound optics.

37. Education.gov.uk, 2010. Education.gov.uk (Image). [Online] Available at: http://www.education.gov.uk [Accessed 03 03 2016].

38. Enfield, 2004. Enefield.eu (Image). [Online] Available at: http://www.enefield.eu [Accessed 02 02 2016].

39. Ener Concept, 2004. Enerconcept (Image). [Online] Available at: http://www.enerconcept.com [Accessed 3 2 2016].

40. ESDB, 2016. Biomass Energy (Image). [Online] Available at: http://www.kids.esdb.bg [Accessed 10 1 2016].

41. European Commission, 2015. Intelligent Energy Europe. [Online] Available at: https://ec.europa.eu/ [Accessed 10 04 2016].

42. European Parliament, 2016. Industry, Research and Energy. [Online] Available at: http://www.europarl.europa.eu

43. European Solar Thermal Industry Federation, 2013. Solar Thermal Markets in Europe - Trends and Market Statistics, Brussels: MT.

44. Fleck, B., Meier, R. & Matovic, M., 2002. A field study of the wind effects on the performance of an unglazed transpired solar collector. s.l.: Solar Energy.

45. Global Solar Thermal Energy Council, 2010. About Global Solar Thermal Energy Council (Image). [Online] Available at: http://www.solarthermalworld.org/ [Accessed 10 01 2016].

46. Go Green Heat Solutions, 2005. Flat Plate Collector (Image). [Online] Available at: http://www.gogreenheatsolutions.co.za [Accessed 3 3 2016].

47. Gov.uk, 2015. National pupil projections: July 2015. [Online] Available at: https://www.gov.uk/government/statistics/national-pupil-projections-trends-in-pupil-numbers-july-2015 [Accessed 02 04 2016].

48. Gov.uk, 2016. Department of Energy and Climate Change. [Online] Available at: https://www.gov.uk/government/organisations/department-of-energy-climate-change

49. Green Building Advisor, 2003. Green Building Advisor (Image). [Online] Available at: http://www.greenbuildingadvisor.com [Accessed 26 11 2015].

50. Greenstream Publishing Limited, 2015. Solar Angle Calculator. [Online] Available at: http://solarelectricityhandbook.com

51. GSTEC Council, 2015. Solar Thermal World - Group (Image). [Online] Available at: http://www.solarthermalworld.org/ [Accessed 2016 02 23].

52. Hall, R., Ogden, R., Elghali, L. & Wang, X., 2011. Transpired Solar Collectors for ventilation air heating. London: Proceedings of the ICE - Energy.

53. Handbook, S. E., 2013. The solar electricity handbook solar angle calculator. [Online] Available at: http://solarelectricityhandbook.com [Accessed 2 2 2016].

54. Health and Safety Executive, 2011. Workplace health, safety and welfare - A short guide for managers. London: Gov.uk.

55. Heinrich, M., 2007. Transpired Solar Collectors - Results of a field trial. Judgeford: BRANZ.

56. Herring, N., 2015. School goes green with solar panels and new programme, Oxford: Oxford Mail.

57. Hindhaugh, E., Blanc, A., Mc Evoy, M. & Plank, R., 1993. Architecture and construction in steel. London: FNS.

58. Hollick, J., 1996. World's largest and tallest solar recladding. s.l.: Renewable Energy.

59. Hub, L. C., 2015. Low Carbon Hub Solar Energy for Schools Scheme, Oxford: Low Carbon Hub.

60. Huil, J. R., 1982. Calculation of Solar Pond Thermal Efficiency with a Diffusely Reflecting Bottom. s.l.: Solar Energy.

61. In Light Solar, 2001. Inlightsolar (Image). [Online] Available at: http://www.inlightsolar.com [Accessed 20 01 2016].

62. Inhabitat, 2004. Inhabitat. [Online] Available at: http://www.inhabitat.com [Accessed 19 03 2016].

63. International Energy Agency, 2009. World Energy Outlook, Paris: OECD/IEA.

64. International Energy Agency, 2012. Technology Roadmap - Solar Heating and Cooling, Paris Cedex: OECD/IEA.

65. International Energy Agency, 2016. About International Energy Agency (Image). [Online] Available at: http://www.iea.org

66. Investopedia, 2016. Business activities. [Online] Available at: http://www.investopedia.com/ [Accessed 1 04 2016].

67. Kreider, J. F. & Kreith, F., 1982. Solar heating and cooling. 2nd ed. New York: McGraw-Hill.

68. Krippner R & T., H., 2000. Architectural aspects of solar techniques - studies on the integration of solar energy systems, Copenhagen: EuroSun.

69. Krippner, R., 2003. Solar technology – from innovative building skin to energy-efficient renovation. Amazon: Solar Architecture.

70. Krippner, R. & Herzog, T., 2000. Architectural aspects of solar techniques. Copenhagen: ISES-Euro Solar Congress.

71. Kutscher, C. F., 1982. Design Approaches for Solar Industrial Process Heat Systems. August ed. New York: SERI Report.

72. Low carbon hub, 2016. Solar heating and cooling (Image). [Online] Available at: http://www.lowcarbonhub.org [Accessed 02 02 2016].

73. Marken, C., 2009. Solar Collectors, New Mexico: Home Power.

74. Matters, E., 2014. Evacuated tube collectors (Image). [Online] Available at: http://www.energymatters.com.au [Accessed 14 3 2016].

75. Mauthner, F. & Weiss, W., 2012. Solar Heat Worldwide, Gleisdorf: Institute for Sustainable Technologies.

76. Mauthner, F. & Weiss, W., 2014. Solar Heating Worldwide, Gleisdorf, Austria: AEE INTEC.

77. Munari-Probst MC, Roecker C & A., S., 2005. Architectural integration of solar thermal collectors: results of a European survey. 1st ed. Amazon: MCA.

78. Munari-Probst, M. & Roecker, C., 2007. Towards an improved architectural quality of Building Integrated Solar Thermal Systems (BIST). s.l.: Solar Energy.

79. Munari-Probst, M. & Roecker, C., 2011. Architectural integration and design of solar thermal systems. London: Routledge.

80. Office for National Statistics, 2015. Families and Households, 2014, London, UK: Office for National Statistics.

81. Oxford Mail, 2015. Mission to turn schools and businesses solar, Oxford: Oxford Mail.

82. Pateman, T., 2015. Rural and urban areas: comparing, London: Inter-Departmental Urban-Rural Definition Group.

83. Person, C. & Anderson, N., 2007. Solar wall monitoring. Bracknell: BSRIA Limited.

84. Power of The Sun, 2010. Glazed flat plate collectors. [Online] Available at: http://www.powerfromthesun.net [Accessed 1 12 2015].

85. Press, C. U., 2016. Camridge Business English Dictionary. Cambridge: Cambridge University Press.

86. Rabl, A. & Nielsen, C. E., 1975. Solar Ponds for Space Heating. 17 ed. New York: Solar Energy 17.

87. Regional Center for Renewable Energy and Energy Efficiency, 2010. About RCREEE. [Online] Available at: http://www.rcreee.org/ [Accessed 02 01 2016].

88. Renewable Energy Focus, 2009. www.renewableenergyfocus.com. [Online] Available at: http://www.renewableenergyfocus.com [Accessed 01 03 2016].

89. Renewable Energy World, 2016. Solar heating and cooling. [Online] Available at: http://www.renewableenergyworld.com [Accessed 14 1 2016].

90. Revolution Power Solar Thermal Heating Limited, 2005. N G Bailey - Edlington High School, South Yorkshire, Newton Aycliffe: Revolution Power Solar Thermal Heating Limited.

91. Revolution Power Solar Thermal Heating, 2006. Mexborough High School, South Yorkshire - Ground surface heat pump and solar, Doncaster: Revolution Power Solar Thermal Heating.

92. Revolution Power, 2004. Revolution Power (Image). [Online] Available at: http://www.revolutionpower.co.uk [Accessed 01 03 2016].

93. Rimstar.org, 2016. Solar energy (Image). [Online] Available at: http://rimstar.org/ [Accessed 2 2 2014].

94. Rowell, A. & Simpson, A., 2016. Guest Blog: How schools can take action on climate change, Oxford: Hub Solar Schools.

95. Sbec, 2003. Sbec. [Online] Available at: http://www.sbec.eu.com [Accessed 09 04 2016].

96. SBED, 2006. sbed.cardiff.ac.uk. [Online] Available at: sbed.cardiff.ac.uk [Accessed 01 2 2016].

97. SCG GRP, 2004. SCG GRP. [Online] Available at: http://www.scg-grp.com [Accessed 20 02 2016].

98. Schools, S., 2016. Thermal energy. [Online] Available at: http://www.softschools.com [Accessed 2 2 2016].

99. Service, Government Statistical, 2015. The 2011 Rural-Urban Classification for, London: Defra Rural Statistics, May 2015.

100. Shanks, R., 2002. BWE Partnership. [Online] Available at: http://bwep.co.uk/ [Accessed 5 01 2016].

101. Shiftnrg, 2004. Shiftnrg. [Online] Available at: http://www.shiftnrg.com [Accessed 09 03 2016].

102. Solar Century, 2015. Eynsham Pavilion goes solar, Oxford: Solar Century.

103. Solar Energy Industries Association, 2011. Solar Heating & Cooling: Energy for a Secure Future, Washington: Solar Energy Industries Association.

104. Solar Panels Plus, 2014. Solar Panels Plus, Pittsboro, NC: Solar Panels Plus.

105. Solar Panel Cleaners, 2004. Solar Panel Cleaners (Image). [Online] Available at: http://www.solar-panel-cleaners.com [Accessed 03 03 2016].

106. Solar Schools, 2010. Solar Schools. [Online] Available at: http://www.solarschools.org.uk [Accessed 05 01 2016].

107. Solar Tubs, 2005. Solartubs (Image). [Online] Available at: http://www.solartubs.com [Accessed 01 04 2016].

108. Solarta, 2004. Solarta (Image). [Online] Available at: http://www.solarta.com [Accessed 04 04 2016].

109. Sources, Centre for Renewable Energy, 2009. High solar fraction heating and cooling systems with combination of innovative components and methods, Attiki: Centre for Renewable Energy Sources.

110.   Tabor, H., 1981. Solar Ponds. 3 ed. New York: Solar Energy 27.

111.   Tata Steel UK Limited, 2012. Integrated solar air heating solutions. 1st ed. London: Colorcoat Renew SCR Brochure.

112.   The Renewable Energy Centre, 2005. The renewable energy centres. [Online] Available at: http://www.therenewableenergycentre.co.uk [Accessed 01 03 2016].

113.   Tutiempo, 2000. Average annual climate values. [Online] Available at: http://www.tutiempo.net [Accessed 5 1 2016].

114.   UK Crown, 2013. Workplace (Health, Safety and Welfare) Regulations. London: Gov.uk.

115.   Volker Quaschning, 2004. Volker-quaschning. [Online] Available at: http://www.volker-quaschning.de [Accessed 09 01 2016].

116.   W Weiss, I. S., 2001. Facade Integration – a new and promising opportunity for thermal solar collectors. Delft: s.n.

117.   Weinberger, H., 1964. The Physics of the Solar Ponds. 45 ed. s.l.: Solar Energy.

118.   Williams, R. & Manager, P., 2016. SBED | Sustainable Building Envelope Demonstration. [Online] Available at: http://sbed.cardiff.ac.uk/ [Accessed 24 02 2016].

119.   Williams, S., 2014. Solar Panel Cleaning on Banbury Academy Completed, Telford: Solar Panel Cleaners.